ART, SCIENCE AND HUMAN PROGRESS

ART, SCIENCE AND HUMAN PROGRESS

The Richard Bradford Trust Lectures
Given Between 1975 and 1978
Under the Auspices of the Royal Institution

Edited by
R. B. McCONNELL

John Murray
Albemarle Street, London

Lectures delivered at the Royal Institution are published in the *Proceedings of the Royal Institution*, and acknowledgements are made to the respective publishers for permission to republish the ones in this book: the lectures by Lord Clark and Professor Kitto appeared in volume 49 (1976) published by Applied Science Publishers Ltd, those by Jacquetta Hawkes, Sir Peter Medawar and Professor Samuel in volume 50 (1977), which, together with Professor Wickham's lecture in volume 51 (1979) was published by Taylor & Francis Ltd, while Sir Ernst Gombrich's lecture in volume 52 (1980) was published by Science Reviews Ltd.

Erratum

On page 84, the sentence beginning "When Edmond Halley . . ." should read ". . . he was doing no more than social scientists . . ."

First published 1983
by John Murray (Publishers) Ltd 50 Albemarle Street, London W1X 4BD

Typeset by Inforum Ltd, Portsmouth
Printed and bound in Great Britain at The Pitman Press, Bath

British Library Cataloguing in Publication Data

Art, science and human progress.
I. McConnell, R.B.
082 PN6014
ISBN 0-7195-4018-6

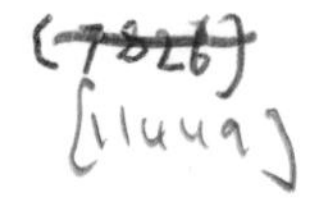

Contents

Illustrations

Illustration Acknowledgements

1, 2, 6, 7, 8, 9, 10, Jacquetta Hawkes; 3, Museum of Fine Arts, Boston; 4, 5, Hirmer Photoarchiv, Munich; 11, National Portrait Gallery; 12, 13, Kunsthistorisches Museum, Vienna; 14, 20, 22, 23, The Louvre; 15, 16, Museo Nazionale, Naples; 17, 18, Her Majesty the Queen, Windsor; 19, British Museum; 21, Metropolitan Museum of Art, Rogers Fund, 1917; 24, 25, 27, 29, National Gallery; 26, 28, Tate Gallery; 30, Royal Institution of South Wales; 31, National Gallery of Victoria, Melbourne

Preface

This volume contains seven lectures endowed by the Richard Bradford Trust. This Trust was founded in 1969 by Dr R. B. McConnell, a professional geologist with a love for the arts, in honour of his great-great-grandfather, the Rev Richard Bradford, who was a pioneer Protestant clergyman in Eastern Canada. These lectures were all delivered in support of the main theme of the Trust which is to study the arts, particularly those of ancient Greece, with a view to assessing the validity of using creative art as a means of examining human behaviour, and to compare this method of investigating the moral, social and spiritual characteristics of our civilisation with the scientific method, as viewed at the present time, which has elucidated so successfully the nature and functioning of the material universe.

All the lectures were given in 1975–78 under the auspices of the Royal Institution, and we wish to thank the Director, Sir George Porter, for his help and encouragement, and his staff for their valuable assistance. We are greatly indebted to the Governors of the Institution for making their premises available to us, because the high prestige of this ancient foundation enabled us to invite outstanding authorities to lecture on subjects germane to the purposes of the Trust.

The Trust is indebted to Professor G. W. A. Dick through whose good offices the Ciba Foundation kindly provided residential accommodation and conference facilities for a symposium under the Chairmanship of Lord Briggs, which followed the lectures. Abstracts of papers discussed at this Symposium are given as an Appendix to this volume.

Dedicated to the memory of
H. D. F. Kitto

INTRODUCTION

Objectives and Background of the Richard Bradford Trust

R.B. McConnell

The Richard Bradford Trust was founded to explore the relationship between the methods of scientific investigation and of artistic creation in literature, the visual arts, and other art forms. The basic reason for its establishment is to test the hypothesis that, although there are wide differences between the sciences and the arts, there is an analogy between the methods practised in these two modes of human activity; and that, as science analyses the material world, so art forms express vividly the human view, not only of the world around us, but also of the essence and purpose of our existence. The scientist observes nature and the material world, and attempts to synthesise his ideas in hypotheses which must then be subjected to experimental verification or falsification in the medium to which they refer. In an analogous manner, the artist or writer synthesises his ideas in a work of art which is then presented to human audiences for approval or disapproval, and the aim of the Trust is to encourage discussion of the proposition that this public exposure to a human audience is the equivalent of the testing of a scientific hypothesis by the experimental method. The differences as well as the analogies between the scientific and artistic approaches are to be discussed. It is emphasised that the judgement of one audience or of one whole generation on a work of art can never be final, and some art which is 'before its time' may eventually be accepted as 'classical', whereas some will be forgotten or remain only as museum pieces.

It is also argued that ancient Greek drama is an excellent example of the artistic method of submitting ideas for approval or rejection by an audience, and thus verifying or falsifying certain modes of human behaviour. Although the 5th-Century Greeks are far from us in time, and some of their mental background is difficult to grasp, their questing attitude towards life is remarkably similar to that of our modern age and will reward objective and sympathetic study. In addition to their literature, the Greek view of life was also outstandingly exemplified in the visual arts. In a similar manner dramatists of later ages have dealt with the common problems and emotions affecting humanity and have depicted the resulting reactions in a manner which may be more effective than direct moral instruction. Shakespeare is an outstanding example of a dramatist mirroring the behaviour patterns of a Western civilisation moulded on the Greco-Roman-Judaic model, influenced by Christian, Romantic and Renaissance ideals. The continuing popularity of his best plays shows how deeply audiences are still moved by the convincing picture of human frailties and virtues which he presents.

An example of a similar great master of the visual arts is Michelangelo, and the generations are never tired of contemplating the beauty and confidence of Youth in his 'David', or the wisdom and compassion of Experience in his 'Moses'. Likewise with music: who has not been moved by Beethoven's visions of beauty and joy?

The methods of science have led to discoveries and inventions which have conferred enormous benefits on mankind when wisely used. As science has enabled us to explore the material world and turn it to our benefit, so the arts should enable us to explore the world of human behaviour. The development of science-based industry has produced a great revolution in the human condition but has brought to light many harmful by-products which should be eliminated. In the same way the display of human behaviour on stage, in novels and in the visual arts is surely essential in guiding the progress of civilisation. Harmful modes of behaviour as shown up for instance in Shakespearean tragedy and Restoration comedy, become exaggerated and cause revulsion and the search for pathways leading to a more equable and inwardly satisfying culture.

The world is now in a period of great ferment and many believe that Western civilisation has failed. It is the purpose of this Trust to initiate discussion of the proposition that the emphasis on materialism during the last three centuries has led to an imbalance in a culture fundamentally based on spiritual values, justice and mutual tolerance and that this deviation could be corrected by a renewed emphasis on art forms.

The growth of science brings us ever closer to visualising the mechanisms and nature of the universe, whereas art presents vividly the contemporary aspects of human society and the problems which it must face. If science helps to explain the material world, then artistic endeavour should help towards establishing a new social structure adapted to the changed modes of life introduced by the progress of medicine, world communications and other by-products of the scientific revolution.

The study of scientific and philosophical thought along modern lines, indicates that a scientific hypothesis can only be correct relative to the frame of reference in which it has been developed, but that it can be effective within that frame, as for example Newton's theory of gravitation was still adequate for the recent Apollo lunar project. If, therefore, the analogy between the scientific and artistic methods developed in this paper is valid, then art forms should be an important means of regulating social and moral codes and adapting them to the modern views of the universe.

Acceptance of this proposition could, it seems, make some contribution towards removing the present-day doubts regarding the direction which our civilisation should now take. As a professional scientist I appreciate to the full the great and predominantly beneficial revolution which the applications of science have brought to the world, but I feel that the material aspect of these has already been greatly over-developed. The material tendency of the experimental phase of science which began with Galileo and was popularised by Francis Bacon, has now come full circle with the discovery of the uncertainty principle and the conception that matter is after all only a particular manifestation of energy. Scientific thought, has in fact advanced so far ahead of moral and social development and placed such powers in human hands that our civilisation is rather like a boy of twelve with a loaded six-shooter in his hands. I suggest that it is through a greater appreciation of the importance of art in developing social ideals that mankind can restore the balance of good sense in human progress.

The scientific method

The scientific method of investigation as seen today, falls into three phases:

1. observation,
2. correlation of facts observed to formulate a working hypothesis,
3. the hypothesis or generalisation must then be subjected to strictly controlled testing aimed at either verifying it or proving it false.

The principles of verification in the sciences of physics and chemistry for instance, in which 'variables' can be quantified, are well understood and also apply to many of the allied sciences dealing with organic and inorganic matter on this earth and throughout the universe, but their application to the behaviour of living beings, and particularly to human problems, is sometimes very sketchy. The accuracy of verification depends largely on the number of 'variables' to be studied and the relationship of the time-lapse of the study to the life-cycle of the phenomenon or of the observer. For instance, the Newtonian laws of gravitation are reducible to a number of variables conveniently accessible to mathematical treatment, and the time-lapse of planetary movements can be evaluated and is small with reference to the observer's life-span. On the other hand psychological or social studies of humanity are bedevilled by the very large number of variables and the short duration of the time-lapse of most investigations compared with the life-cycles of the generations which may be influenced by the particular phenomenon studied. The strictly 'scientific' value of many psychological and social studies may thus be questioned, owing to the absence of adequate methods of testing.

The scientific method does not result in the discovery of *absolute* laws of nature, but in approximations which are related to the frame of reference of the investigation. Thus, Newtonian physics has been modified by Einsteinian relativity theory, but was sufficiently accurate to serve as a basis for all aspects of the highly successful Apollo lunar exploration programme. On the other hand, the estimates of the age of the earth by 19th-Century physicists were much too short by comparison with those based on geological evidence because one extremely important variable was overlooked, namely the energy released by the natural radioactive decomposition of minerals in the earth's mantle and crust. A similar advantage of geological observation and deduction over the more quantitative physical sciences became apparent during this century. In the 1920s many geologists accepted the then newly-proposed theory of 'continental drift', but this soon became unpopular because it did not conform to the contemporary physicists' concept of the earth's structure. In the last twenty years however, the conception of mobilism has become generally accepted, as new evidence concerning global structure and dynamics has emerged, and a theory of 'plate tectonics' has transformed our global outlook. Thus, a whole generation of geologists was deprived of the advantage of a unified dynamic view of the earth's evolution by the desire for a mathematical accuracy which was in fact misleading because important facts were still unknown. These examples show the value of a scientific method which allows the *averaging out* of the effects of many more variables than can be considered in mathematically-based sciences and also indicates how reasoning analogous to the scientific method may be effective in portraying human behaviour in a manner which describes in the well-known metaphor, the shape and characteristics of the wood without the necessity for a long, detailed, and possibly misleading study of the individual trees.

The great philosophical lessons to be learnt from the progress of science and which the Trust was founded to explore and discuss appear to be:

1. That absolute truth is unobtainable by us; as Karl Popper has shown, we can only achieve 'conjectural knowledge'. The relative 'truth' of any proposition is entirely dependent on the frame of reference under which the investigation has been made and may be largely modified by the discovery of an unsuspected variable, such as for instance, radioactivity or the fourth dimension.

2. That science, far from destroying the beauty and romance of the world as seen by artists, musicians and writers, enhances it by revealing the underlying reasons and purposes. The origin of our world and the universe in which it is placed is essentially a mystery about which man cannot abstain from speculating, but it has been said: . . . 'by their fruits shall ye know them'. So, if we want to approach any solution of the mystery, surely we must study the great hidden beauty of the mechanisms not only of the material world but of life and above all of human history and evolution.

3. To my mind however, the most valuable lesson of science is that there is a *method* by which we can establish aspects of truth, which, although only relative to our own frame

of reference, are greatly satisfying. Keats believed earnestly in the truth of the imagination, but says in one of his letters: . . . 'if only we could *know*'. Perhaps we can never *really* know, but if my proposition of the analogy between the methods of science and art is at all sensible, then we can look forward to attaining a very satisfying approximation to truth. To test this proposition is the fundamental object of the Trust.

The essence of the research problems which the Trust was established to study, lies in the proposition that art forms in the widest sense, constitute a method for investigating the nature of our world and in particular of human behaviour, which is analogous to that of a true scientific investigation, and also involves a method of verification. The importance of the study of philosophy, psychology and social behaviour is recognised, but the object of the Trust implies that researchers should investigate the proposition that the methods of *verification*, inherent in science and in true art forms, are more effective. They should ask whether artists, poets, novelists and particularly dramatists, not only 'hold the mirror up to nature' but can help greatly in the development of social trends tending to increase the stature of humanity in the world and thus continue the long process of the evolution of consciousness which dates back at least to the origin of the genus *Homo*, perhaps as much as five million years ago. It is quite true that certain artistic manifestations may seem to be demoralising, but these should be seen against the background of evolution over the centuries as will be shown in a later section.

It is suggested that the three fundamental steps of the scientific method, as outlined above, namely, observation, generalisation and testing, are also followed in the development of any work of art. The artist or writer must obviously be an intent observer and it is well-known that, as in the case of the scientist, a retentive memory is essential. As the observed facts and images build up in the artist's mind, probably to a large extent subconsciously, he experiences a desire or compulsion to simplify the picture by expressing a generalisation which must, in view of the immense number of variables involved, be largely the result of intuition and, as the scientist formulates a hypothesis, so the artist expresses his ideas in the form of a picture, sculpture, book, play, or in musical form. But the process does not stop there and the artist feels the necessity of displaying his work just as the scientist aims at publishing his hypothesis. In the case of the scientist publication is aimed at arousing discussion and wider study which may result in either falsification or verification; and my proposition is that the exhibition, book, play or music is equally, although perhaps unconsciously, aimed at acceptance by a human audience and hence *verification* of the aritst's ideas within the frame of reference in which they were conceived.

The first two stages of the process just described are common not only to science but also to other disciplines, such as philosophy, psychology, social studies and so forth. I suggest, however, that the third process, verification or falsification, can only be effectively applied in science and in art. In the case of philosophy propositions are generally dependent on intricate logical argument, frequently refutable by specialised intellects, and the variables are many and difficult to define. Psychology must take into consideration a number of variables very large in proportion to the methods of investigation available and the time-lapse necessary for verification is also large, or very large with regard to the lifespan of the investigator. Social studies are subject to an even greater extent to the same limitations as psychology. But a work of art is actually subject to verification or falsification. This whole problem is one which should be further studied under the terms of the Trust.

As we have seen earlier, a scientific 'truth' can only be acceptable within the frame of reference in which it was developed. In the same way, a picture, a poem, a symphony or a play is developed in a human brain as the result of human observation and must be judged on its apparent 'truth' relative to the human frame of reference. So when we look at a picture, watch a play or listen to music, we like it if it 'rings a bell', that is to say, if we feel it expresses some way of looking at a subject which we feel is true, though we may not consciously have recognised this truth before. The artist's view is thus either verified or rejected in the frame of reference to which it relates, namely, human experience. The verification is, of course, not merely limited to our own view, or to that of contemporary public opinion, but must be agreed or rejected over the ages by a large majority of those to whom it is submitted and by successive generations. Thus, even if the work of art is 'ahead of its time', and generally rejected by contemporary opinion, future generations may recognise its truth and it may thus attain verification. On the other hand, more frequently, it may simply come to be regarded as an irrelevant period piece.

I feel that I can speak with some familiarity about the scientific method because as long ago as 1921 I became a pupil at the University of Lausanne of a remarkable geologist, Maurice Lugeon, universally recognised as the principal architect of the revolutionary conception of the geological structure of the Alps, which has had such an enormous influence on our science. I was introduced by him, in actual field work, to a method of investigation which consisted in development of a series of hypotheses which could be tested in the field and either falsified or verified, a method now well known through the works of Karl Popper, but used many decades earlier, I am quite sure, by many practising scientists. Experience with this method in later years has convinced me that it is the most effective way of advancing scientific knowledge.

Imagination and reasoning in science and art

I do not think the overall scientific approach has changed

much but it is perhaps true to say that philosophical understanding of these methods has changed. Briefly, the method of investigation which has developed modern science, as described for instance by Popper, Bronowski and Medawar, is no longer the reasoned step by step induction process which we associate with Bacon. It actually involves rather a close collaboration between the imaginative and analytic faculties of the brain. It is thus much closer to the act of artistic creation than is generally recognised. I think that it can be described as falling roughly into three stages. First, an assiduous, or far-reaching collection of facts. Now the actual facts appertaining to any geological problem, for instance, are far too numerous to be retained in the conscious memory, and must somehow be stored subconsciously. As the investigation proceeds, the second step is that the information acquired must be synthesised to form a hypothesis. This process may be compared with the working of a computer, except that the facts or 'variables', are far too numerous to be dealt with by even the most modern computers. The third, and most important step, is finding a way of testing or cross-checking the hypothesis, either to verify or falsify it, preferably by several different methods.

I must make it quite clear that I regard the disciplines of the sciences and of the arts as in different categories, but what I want to suggest is that the first two steps which I have described above must also, consciously or unconsciously, be taken by the creative artist, be he painter or playwright or whatever. Observation is a continuous process throughout life, and from the moment when a baby's eyes first open he or she must be observing the surroundings and behaviour of people. The retention of impressions received in early childhood has been commented on by Colin Blakemore in a recent television series on the brain; Chomsky has stressed the integral relationship between language and the human mind. With advance of years accumulated information must form a vast store somewhere in the conscious or subconscious mind. Moreover, I suggest that artists would not be artists unless they were intensely observant and endowed with a retentive memory, as is the case with scientists. Here may I borrow from Professor Kitto's abstract (see Appendix, page 180) to illustrate what I mean? He says:

> It seems to have been instinctive in the Greeks to seek the general in, or behind, the particular. Is that the reason why they contributed more than any other ancient peoples to the development of modern science?

Here Professor Kitto has clearly brought art and science together. Is not this a good description of the process of inspiration which drives the artist or writer, as well as the scientist? Can it not be agreed that a poem, a play or a picture is a generalisation from the vast accumulation of particulars which an individual has stored in his conscious or subconscious memory?

The third step in the scientific method – verification or falsification – may be applied to the work of the artist, in the appropriate frame of reference, by submitting it to a human audience; and by averaging out their judgments over the generations. This is a controversial problem which I hope will be treated in future discussions.

As already stressed, one of the greatest contributions of science to our culture appears to be a special new conception of knowledge built up by the scientific method of multiple hypotheses, continually cross-checked and tested in the medium in which they were conceived. Medawar in his book *The Hope of Progress* (1972), says: 'One of the very greatest of all discoveries was the connection between imagination and reasoning, between the inventive and critical faculties'. This underlines the value of the new picture of the scientific method. New conceptions are built up whose limits of reliability can be estimated. However confident 19th-Century scientists may have been that absolute truth was just around the corner, present day scientists know that they can never attain more than an approximation to truth, and that their conclusions can only apply within the medium in which they are working. They are always aware that theories may be modified or contradicted by the discovery of previously unknown variables. Sometimes, however, they have the great satisfaction of feeling that a hypothesis has stood up to tests, and to the impact of new information and may therefore serve as a basis for further work. If the analogy between the artistic and scientific methods is at all valid, then by transposition from one quite different field to another, the artistic method can also attain what Popper has called 'conjectural truth'.

A corollary of what I have been saying is that to bridge the gap between the 'two cultures' of Lord Snow, it is necessary for the traditional culture to appreciate the new scientific way of thought, so well described by Popper and Medawar, and for scientists to cultivate knowledge of the artistic and literary tendencies of their time.

There is another important characteristic which I suggest is common to science and art. Scientists deal with observed facts and must be very strictly objective over the evidence they use: without a high standard of integrity their work is not only useless but may be damaging to future investigations. Art deals in a more subjective way with the facts of nature and human life, but it is generally conceded that it must also have a high degree of integrity – the phrase 'it rings true' or the opposite, is constantly met with in art appreciation.

Verification of relative truth in art forms

Works of art are the products of the human mind, formed probably by a subconscious inspiration which synthesises, in the manner of a super-computer, a vast number of impressions stored in the artist's memory. Thus, when the work is displayed to a human audience, an immediate reaction can be obtained which is also in part instinctive or subconscious. This can average out the effect of the huge number of variables involved. The reaction will vary considerably according to the prevailing *zeitgeist*: the reaction in the 'age of reason' would thus have been very different from that of the present age, and so forth. Think of the shelves of books which have been generated by a single play such as *Hamlet*. The play obviously rings true, but in so many ways that each generation sees it slightly differently and it is hard to imagine the discussion ever dying out. Verification by exposure to human experience can extend over succeeding generations, and so with the passage of time, certain artists and periods of art come to be regarded as of major importance; in other words, as depicting outstandingly constructive lines of human development.

Social philosophy has dealt much with the concepts of good and evil, but throughout history, periods have arisen in which new generations have questioned contemporary value judgments and vital discussions have ensued. Such crises occurred, for instance, in the age of Socrates and Plato, at the Reformation, and one is occurring at the present time. If we are to approach these problems reasonably, we have no right to make preconceived value judgments, and the younger generation is justified in questioning accepted standards. Any hypothesis dealing with good and evil must be subject to a process of verification or falsification averaged out over the generations. I suggest that this is precisely where the artistic method is of greatest value.

It has been hinted above that the question of whether some artistic manifestations were demoralising should be viewed in the perspective of evolution. The principle of natural selection enunciated by both Darwin and Wallace, is now very widely accepted as being one of the chief factors in directing biological evolution, and it was largely inspired by the ideas of Malthus. Thus, the principle of the 'survival of the fittest' is regarded as vital in the mechanism by which life forms have evolved into more and more complicated entities capable of exploiting new life support patterns.

Variations or sports which are poorly adapted or retrogressive are naturally eliminated in subsequent generations, whereas those which are 'fit' survive. Comparison of views on art and science thus suggests that an important criterion of *fitness in cultures* is in fact *survival*. It appears, for example, that the concept of good and evil may have arisen through the recognition, perhaps in part unconscious, of the survival value of certain modes of behaviour. This could have taken place during the many generations since the genus *Homo* began to form social groups to profit from the survival value of co-operation in hunting and agriculture. Thus, behaviour which benefited the community came to be regarded as 'good' and was encouraged by elders and tribal leaders, whereas that which damaged it was 'evil' and subject to punishment. Such a proposition needs much elaboration, but acceptance of the general idea would lead to the conclusion that the facts of history can help towards a comparative study of good and evil or, in the modern spirit, of 'better' and 'less good'.

Averaging out of variables in art forms

As we have seen that the 'averaging out of variables' is an important factor not only in testing hypotheses in the biological and geological sciences, but also in verifying the truth of ideas expressed in works of art, so we can understand why artists of universal and ageless importance are very dependent on the period in which they chance to be born. What I would term the 'sounding board effect' comes into play if the artist is expressing ideas which his contemporaries appreciate, and his stature and influence will hence be greatly increased. Who can doubt that the rapturous acclaims of Homer's audiences encouraged the poet in his descriptions of Greek life to stress opinions on what he considered to be good, less good, or evil? He influenced generation upon generation of Greeks and his influence is still alive with us today.

In the great period of Athens, sadly short as Thucydides has shown, the dramatists were able to test their profound interpretation of the deepest facets of human behaviour directly on a great sounding board, namely the audiences of thousands of eager citizens who attended the performances at the Dionysian festival. This highly intelligent and lively gathering would tell the dramatist at once whether what he was saying was true or questionable as they saw it in the human medium. Thus, so many of the principles on which our civilisation is based – the evil in revenge and the necessity of impartial justice; human stature and respect for the individual; the evil in pride; the evil of yielding to the emotions; and many others – were aired and found to be essential factors without whose recognition large societies could not live in peace. To a lively and intelligent public, eager to probe to the bottom of any statement, it is not enough to 'preach' social behaviour any more than for a scientist to express his hypothesis in a simple proposition. Fortunately, both art and science have direct methods for the verification of ideas, some rapid and others requiring testing over many generations. Thus, social ideas, if put forward in the form of art, and particularly in the theatre, can be tested directly against the sounding board of the human beings who form its frame of reference.

Drama as a means of analysing the human situation

Looking back at the development of civilisation, Greek literature occupies a specially significant place, and, although my knowledge of it is very superficial, I consider that from the point of view of this investigation, Greek drama of the 5th Century B.C. can be accepted as an outstanding example of the application of the artistic method to the study of human behaviour. Although the meaning of Greek tragic drama has been differently interpreted, we accept the view of our lecturer H.D.F. Kitto that in Athens drama was recognised as a potent method of analysing the elements of moral and social behaviour before an audience which would not tolerate preaching, but was ready to be influenced by what the great poets had to say about the fundamental problems of life. The importance attached to drama by the citizens is shown by the time and labour lavished on the annual festival of the Dionysia. With hindsight one can say that they were exploring modes of moral behaviour by the artistic method of verification or falsification, as I have already proposed. I think it is very likely that the people, as suggested by many writers, and more or less confirmed in *The Frogs* of Aristophanes, did actually realise that the dramatists were expressing new moral outlooks, but they were only prepared to accept the plays which rang true. The citizen audience was thus acting as a sounding board, reinforcing what they approved and condemning what they regarded as false. History has shown, alas, that the social ideals developed in Athens during the early 5th Century did not stand up to the stresses of the disastrous Peloponnesian war, but the fact that the plays we now study survived through the ages proves their continued popular appeal.

Shakespeare also lived in a period in which his poetry was appreciated, and it is difficult to exaggerate the influence of his plays on human outlook and behaviour, whereas the contemporary blood-and-thunder Elizabethan drama, although more popular at the time, is now generally regarded merely as a curiosity. It should always be remembered that both the great Greek dramatists and Shakespeare were good 'box office' in their day. They also qualify as examples of 'fitness', if we accept the law of the survival of the fittest. Ancient texts such as the Psalms of David, Homer's epics, and the Greek dramas have only come down to us because generation after generation decided that they were worth recopying and circulating, and thus the chances of survival of a manuscript depended on the popularity of the writing.

As an example of the way in which art may not only

mirror but influence human behaviour, let us look at the *Oresteia* of Aeschylus, the great revenge problem story of Agamemnon and Orestes, and compare it with Shakespeare's *Hamlet*. I must emphasise my indebtedness to Professor Kitto for the comments in his books both on the *Oresteia* and on *Hamlet – Form and Meaning in Drama* (1956), *Poiesis: Structure and Thought* (1966). Although my original conception was formed long before I came across his books, my present views cannot escape being tinged to a large extent by Kitto's perceptive writings. I suggest that Shakespeare deliberately adopted the Agamemnon-Orestes theme in order to present his views on this age-old revenge myth. Homer first tells quite simply the story of Orestes, who, as soon as he attained manhood, acquired fame by avenging the murder of his father. Aeschylus expands the story into a trilogy of plays, telling first of the murder of Agamemnon by Clytemnestra and Aegisthus, and then the revenge by Orestes. In the third remarkable play, the *Eumenides*, we are shown Orestes pursued relentlessly by the Furies for the murder of his mother. Finally, he is brought to trial in Athens in a scene in which he is judged by a human jury, and acquitted by a ruling of the presiding Goddess Athena, because the votes for guilt and innocence were equal. Incidentally, do not we have here a first conception of the humane notion now accepted in English law that a person is innocent until proved guilty? Athena then persuades the Furies to settle in Athens as the Eumenides, or beneficent ones, to ensure that crimes are brought to light and treated with justice and compassion. Surely this marks a fundamental revolution in human thought.

Sophocles, the conservative, takes up the story again, but does not introduce punishment for Orestes, because he believes, as Professor Kitto explains, that justice, the Greek *dike*, has merely been accomplished in the traditional way by the killing of the two murderers, and thus a crime has been expiated by filial piety, and the even course of nature restored. Euripides, the great liberal, influenced by the tragic years of the Peloponnesian war, returned to the story several times, stressing the horror of matricide, and only allowing Orestes to be pardoned after long years of persecution by the Furies.

I suggest that Shakespeare decided to give his own version of this famous revenge story. Approaching it as a Renaissance writer, influenced by a thousand years of Christianity, he realised that he would violate the spirit of his age if he allowed a man of Hamlet's humanity to kill his mother. Therefore Hamlet spoke daggers to her but used none. It seems relevant to my theme of the interrelation of art and human behaviour, that this fundamental problem has been treated, and differently interpreted, by great poets throughout history, and that the Renaissance version should receive verification through the acclaim of *Hamlet* by so many generations. The old Greek versions are still held in high honour, and I suspect that the *Oresteia* trilogy could, through the introduction of the human jury, as well as through the birth of compassion, and the concept of the eventual forgiveness of crimes, come to be regarded as an important landmark in the history of social progress – a further illustration of the development in Greece of that 'respect for the individual' which is the hallmark of Western civilisation.

Another notable example of the contribution of art to civilisation is the generalised and poetic vision of evolution given by John Keats in 1820 in the unfinished poem *Hyperion*, Book 2, lines 165–243. Keats abandoned this poem because it was criticised as too reminiscent of Milton; the language may be but the philosophy is certainly not, and it could be regarded as a 'poem for today'. Superficially this reference to Keats may not appear relevant, but I attach great importance to it as I think that we are moving to a greater appreciation of the meaning of art. In Keats' time the poets of the romantic revival were swamped by the rise of industrialism as development turned to the 'dark satanic mills' of William Blake. Keats was regarded as simply a poet

– which he undoubtedly was – but study of his letters shows a capacity for deep thought. His acceptance of the controversial notion of evolution as early as 1820 is striking. It is only then that science realised that the dominant animal populations once consisted of primitive dinosaurs, which were succeeded by birds and numerous mammals, including beautiful antelopes, and eventually man. Beauty is frequently associated with efficiency, not only in animals but even in mathematical concepts. Beauty is also perceived in landscapes and buildings, and it is more and more realised that the development of human behaviour is influenced by environment. In view of the present reaction against materialism, we need to turn to a greater practical appreciation of art as an important factor in human progress, and the philosophy of Keats may come to be an important influence. This is a movement which the Richard Bradford Trust would wish to support strongly.

The examples given here are taken from Hebrew, Greek and English art during periods which are outstanding in the history of human development, and their art mirrored this activity. Other great periods of art such as the Romanesque in France and the Renaissance in Italy also accompanied exceptionally fruitful phases of the human spirit, and many examples could be cited of inspiring works from the great masters of the visual arts such as Michelangelo and Raphael, or from musicians such as Bach and Beethoven.

It may be suggested that 'modern art' has been largely affected by the growing materialism of Western civilisation. It appears to me, however, that today there is a possibility that we may be emerging from a long period of materialism and that a realisation of the seriousness and importance of the artistic approach to moral problems may help us to find new ways forward in the development of those mysterious tendencies of the human spirit which, during a few million years, have brought us out of the darkness of the animal world into the light of consciousness, self-control and awareness of social responsibility.

R. B. McConnell

As the energy and imagination of the Western peoples has immensely altered the material conditions of civilisation, so it can also investigate the different paths of moral and political progress which lie ahead by stimulating the dramatic and other artistic methods of presenting ideas to be verified or falsified against the sounding board of the instinctive wisdom of public opinion. The broadminded and tolerant Western civilisation, with its perpetually questing and independent spirit, could thus be rendered more alive and adapted to our much expanded knowledge of nature and the universe.

Ultimate objectives of the Trust

The ultimate object of the Trust is to examine the proposition that the way forward for Western civilisation lies not merely in a repetition of age-old precepts, but in developing the artistic and particularly the dramatic method of verifying or denying their truth and adaptability to the rapidly changing conditions of the modern world. Scientific discovery and technological progress have brought about, within the last hundred years, a completely new concept of man's relation to the universe. At first a wave of self-confidence ensued, and it was felt that 'man had conquered nature': but now, as we perceive more clearly the enormous complexity of the universe, the remarkable evolution of life forms through more than three billion years, and the immense social problems facing our generation, a new sense of humility is inevitable, and is manifested by a proliferation of cults and religions. Scientific practice, however, shows that research in any subject can only be fruitful if pursued within a defined frame of reference and subjected constantly to testing. Therefore, since our problem is the establishment of revised modes of social behaviour to adapt to a new world threatened by over-population, by technical innovation, by ideological and racial conflict, research within a human frame of reference is surely indicated. I therefore suggest that we take a lesson from the ancient Greeks, the founders of the democratic way of life which has enabled the establishment of so many of the personal freedoms we so much prize. They recognised art in all its forms as a weapon in the development of new moral and social codes, a weapon at least as powerful as philosophy, psychology or social science, but which could be constantly subjected to human acceptance or rejection.

While pondering the aims of the Richard Bradford Trust, as set out in this introduction, the writer became aware of grave deficiencies in his reading. In partial remedy for this condition the Royal Institution of Great Britain kindly undertook to arrange lectures on relevant subjects by outstanding authorities. The texts of these lectures form the following chapters.

The relevance of the Trust of the lectures delivered at the Royal Institution

It was indeed an encouraging omen for the Trust when Lord Clark agreed to give the opening lecture, particularly as he chose 'Television' for his subject. He can speak with an unrivalled authority on this subject, and himself set a new standard with his series on *Civilisation* in 1969. From the time in 1937 when he broadcast the first art programme ever from Alexandra Palace, he has believed in the future of television. Since, as he points out, one of its great achievements is to widen people's horizons enormously, with nature films, drama and the other humanities, his talk seems specifically designed to show how television can spread knowledge of the subjects with which the Trust is most concerned – art, science and human behaviour – to an audience of almost unbelievable proportions. People can also form some visual impression of those personalities who are guiding their destinities, and discussion programmes are important because opposing views can be argued before the public.

The lecture by Professor H.D.F. Kitto is most relevant to our Trust as, spiced with his dry wit, he explains why Greek drama plays such an important part in our culture. By illustrating Greek 'wholeness', he shows why their way of life is so relevant to a period such as ours, which is searching for a fundamental meaning in life. Drawn instinctively, as we may be, towards the beauty of Greek mythology, the heirarchy of Greek gods may tend to shock us, holding as most of us do, to the power of a supreme God. Kitto shows, however, how the picture of life as we know it, with its good and evil, is well illustrated by the interplay of the divine and the secular. The passions with which we are created – love, anger, violence, ambition and the rest – can be illustrated by mythical personalities, who, while immortal, can display mortal virtues and weaknesses within a 'whole' ruled ultimately by divine justice.

A point of great importance to us is that Professor Kitto quotes the philosopher A.N. Whitehead, as writing that:

> Insofar as modern scientific thinking has Greek ancestry, its Greek ancestors are not the Greek philosophers but the Greek tragic poets.

Professor Kitto gives reasons for agreeing with Whitehead, and I suggest that his argument calls for careful consideration.

We were fortunate indeed in persuading Jacquetta Hawkes to give the third Richard Bradford Lecture, because, as Professor Kitto has succeeded in entering into the mind of the Ancient Greeks, so she presents us with fascinating thumb-nail sketches of the fourth-millennium Sumerian, Egyptian, and the somewhat later Minoan, civilisations upon which the Western World is founded; and, in contrast has given a glimpse of a dark way of life in Central America which sets off the beauty of the developments in the Middle East. As Jacquetta Hawkes in her outstanding book *A Land* (1951) has shown how life in our islands rises from the rocks, so she also considers the geographic background of the two ancient cultures of the Middle East and suggests that their differing characters may be due to ecological differences between the land of the Nile and that of Mesopotamia.

It will be apparent from the introduction to this series of lectures that the notion of a 'scientific method' for the investigation of nature adopted by the Richard Bradford Trust is based largely on the views of Sir Karl Popper, and it is a great advantage therefore, that we are able to include in this volume a talk by Professor Sir Peter Medawar on 'The Philosophy of Karl Popper', as there is no better authority at the present time on the works of this great philosopher. Sir Peter describes Popper's philosophy under two headings: the Philosophy of History and Social Science, and the Philosophy of Natural Science; and he stresses the importance of Popper's great work: *The Open Society and its Enemies* (fifth edition, 1974), in which a dramatic attack is made not only on the totalitarian theories of Marx, but also on those of Plato and Hegel. Sir Peter concludes that what the philosophy of history and of social studies and the philosophy of natural science have in common is the recognition that human schemes of thought are fallible, and we must proceed by identifying and learning from our mistakes. Thus, he does in fact see an analogy between the creative methods of the arts and of science.

After a good deal of discussion about the application of the scientific method, we are happy to be able to include a paper on 'Some Facts and Theories regarding Research in the Brain': an account of its practical application in research into the neurosciences by Dr. David Samuel, Professor of Physical Chemistry and Head of the Chemistry of the Brain and Behaviour Group, in the Isotope Department of the prestigious Weizmann Institute in Israel. The first problem tackled by Professor Samuel is that of Memory, and he gives a fascinating review of previous research on human memory, followed by an all too brief consideration of a biological hypothesis of its nature. Finally, there is a long consideration of the problem of the breakdown of normal mental functions, and hope is held out of eventual pharmacological treatment of these various disorders.

It would be impossible to find a better authority than Professor Glynne Wickham to give an overview of English drama from the early mediaeval Cycles of Mystery Plays and Moralities. Starting with Hamlet's well-known advice to the players, he analyses the subtle meanings Shakespeare attaches to the words 'mirror' and 'nature', which we tend to overlook, and shows how these were adapted to the political background of those troubled times. He then traces the evolution of drama through the didactic period when it served the Church of Rome up to the coming of the Reformation. With the advent of Puritan Government, the theatre was nearly killed by the narrow views of humourless men, but survived thanks to royal intervention and the good sense of people who understood the abiding significance of the stage happenings. In the 17th Century the advance of science encouraged the codifying of drama into tragedy and comedy; the former analysing the passions, and the latter simply depicting contemporary manners and the difficult situations to which they could lead.

Professor Wickham discusses the competition of Television and the Cinema. He is very optimistic about the future of drama, and points out that we are rediscovering the value of direct dialogue between the actor and his public which promotes a sharing of ideas and the realisation that serious comments must be combined with recreation.

Finally in the last lecture, which he entitled: *Experiment and Experience in the Arts*, the noted art critic, Sir Ernst Gombrich has given a most exciting description of the progression of art, and particularly of painting, from the earliest days. He starts with a quotation from a lecture given in 1836 in this same lecture hall of the Royal Institution, in which Constable claimed painting to be '. . . a branch of natural philosophy'. Sir Ernst quotes from previous lectures in this series, and points out the differences between science and the arts, and the danger of tempting artists to see themselves in the role of prophets and oracles. He gives an excellent account of the reasons for the development of modern art, but shows that it is susceptible to brainwashing and the starting of bandwaggons. However, Sir Ernst is encouraged by the belief that today the young have come round to the realisation that the present 'unprecedented variety of modes, media and effects' can be accepted or rejected: a view which the Richard Bradford Trust attempts to promote.

LECTURE I

Television

Kenneth Clark

It is a great honour to be invited to give the first Richard Bradford Lecture. The aim of the lectures was stated quite categorically by the Richard Bradford Trust. They were to be concerned with the Influence of the Arts and of Scientific Thought on Human Progress. It seemed to me that television qualified under both these heads. I am not qualified to comment on the scientific element of television. I asked our head engineer at the ITA how long it would take me, with a fair grasp of mathematics, to acquire a rudimentary knowledge of the technical side of television. He replied, 'If you really applied yourself, about three years'. As I had been appointed for only two years, I gave up the attempt. But I believe I have some qualifications for concerning myself here with the Art of Television, and speculating on its effects on human progress. I was therefore much relieved when Dr McConnell (the Reverend Richard Bradford's great-great-grandson), the man actually responsible for this Trust, wrote to me, approving of the subject.

That television, either as author, performer or organiser, has been my work for many years is possibly known to some of my readers. I did the first art programme ever done on television in the Alexandra Palace in 1937. It can have been seen by only a few hundred people. Nevertheless, I believed in the future of television. We in England had a start over the rest of the world, and under the enlightened guidance of the BBC I thought we could do an immense service to the arts, the sciences and entertainment. The war put an end to all this, and when British TV was free to start up again America had surpassed it in resources, although America had never equalled England, either in overall quality of performance or in technical competence.

At that time the BBC had a monopoly. I am sure that they used it conscientiously. They were then a most conscientious body of men, although slightly out of touch with public feeling. But monopolies are open to subtle forms of temptation. Before 1952 the leading lady in the 'Play of the Week' was paid £25. And apart from this kind of frugality, monopolies develop a feeling that they are always right, and that new forms of development are out of the question. So when the time came for the Government to consider an alternative service, paid for by advertisements instead of licence fees, I was very much in two minds. I admired the BBC tradition of giving people something better than what they wanted. On the other hand, I thought that a little competition would do no harm. It never occurred to me that I would be asked to become the first chairman of the ITA, and in fact to set it up from scratch – no Director General, no premises, only one or two members of a board rather oddly chosen. I was invited because it was thought that my previous harmless record would disarm criticism. I accepted because I thought it would interest me. There was also a political reason for my appointment, of which I was not aware. Then, as now, it was felt in government circles that the BBC was too far to the left – too 'bolshie' was the word. It was assumed that I would be a Tory party stooge. This turned out to be a miscalculation, as became clear when I appointed as our first Director General Robert Fraser, a former Labour candidate. It is he, more than anyone, who should take credit for the form of the ITA. But I was equally responsible for the choice of our first companies, and for the setting up of 'Independent Television News'. I mention all this only to justify my selection as a lecturer on the subject of television by reminding the reader that I have seen the medium from both sides – or rather, from three sides, as the organiser, the performer and the viewer.

As a performer, I had done just under a hundred programmes. Of course I could not do programmes while I was still Chairman of the ITA. My first programme was an absolute disaster. I had not learnt that what people want is information, not speculative ideas. I had not learnt that every programme must be scripted down to the last word. Arthur Askey may say that he enjoys ad libbing, but even he has cards held up all round to bring him back to the script. After my first fiasco (which wasn't altogether my fault) my contract was very nearly cancelled; but I went on to make about fifty programmes for ATV.

My third qualification is that I see a lot of television. As a rule intellectuals, or near-intellectuals, do not look at television. Most of them do not have a television set in their rooms, and have to go down to the kitchen or the porter's lodge if they want to see a programme in which one of their friends is appearing. I watch it every night, partly because my wife is an invalid, partly because the technical side of the medium interests me, and partly because, in fact, I see a lot of very good programmes.

Having put forward a few qualifications, let me make a frontal attack on my problem. I will begin by saying something so obvious that it is often forgotten. Television is a very expensive medium. When we were trying to find companies to whom we could give a concession – and at that time we wanted only four – we had to tell them that they must be prepared to put down at least three million pounds. I was opposed to the idea that Roy Thompson, who already owned the *Scotsman*, should also own the Scottish TV concession (not that Lord Thompson ever interferes in the policies of his many holdings); and so I went round like a commercial traveller trying to drum up support among my countrymen. I think I got about £60,000. We are a prudent people. How they must have regretted their prudence five years later. Today a company can spend close on a million pounds on a single series. There is no use thinking that you can have on TV the equivalent of a small theatre group, or books published in a limited edition. Even opera, which is notoriously expensive, costs much less; and it is always possible to produce distinguished operas on a small scale, as used to be done, for example, in Aldeburgh. This means that TV must, in the main, appeal to very large audiences, and the number of people who watch – what are known as the ratings – is vital, to commercial TV because of the advertising rates, to the BBC on account of prestige. I should add that the ratings are remarkably reliable, and those who think that they are superior to their fellow men should study them. One finds that one switches off a programme at exactly the same moment as about three million other people. Well, this popular element in TV is a fact and we have got to live with it. But, as I hope to show, it isn't quite as bad as one would think.

The second obvious thing to say about TV is the amazing speed with which it has grown into a sort of necessity. The chief reason is, I think, that it takes place in our homes. This acts in two ways. It can be seen by a whole family and be the subject of comment and discussion. Or it can relieve solitude. A large number of people are lonely, and a number are invalids, and to have someone in their rooms, talking to *them*, in an ingratiating manner, is a comfort to them. To go out alone to a cinema only increases the sense of isolation. One is cut off from it. Whereas, when the same film enters one's room, one is absorbed by it.

Given this element of personal appeal, it is extraordinary how long it took politicians to recognise the value of television. The fact is that for centuries they had shouted at people from outside, which they enjoyed, and they were unwilling to learn the delicate art of talking to individuals from inside – when a lie is more easily detected. Hitler lived before the age of television. We see him often enough, God knows, but he is always barking at a mass rally. Imagine what this extraordinary genius who, by all accounts, could charm as well as intimidate, would have done if he had mastered the technique of direct speech to the individual. Living in an age of mediocrities we haven't begun to realise what the visible appearance of a great spell-binder could do. (Mr Macmillan, in this country, is the nearest we have come to a master of political television. But the only public man I have seen who could combine persuasion with a real sense of authority was General de Gaulle.)

To give an idea of how recently television has emerged, let me mention that when I was making a series called *Civilisation*, the Director of the BBC came to see one of my last programmes (the one about Turner) and said, 'This is no good; it depends on colour, and nobody has a colour television set.' Only seven years ago. So that talking about television is not at all like talking about the effects of the telephone, or the gramophone, or steel and concrete buildings, or the internal combustion engine. To the historian so short a stretch is always rather daunting. However, I do not myself believe that the character of television is likely to change very greatly, and some of its effects are already well enough established to be worth considering.

There are those, like Malcolm Muggeridge, who believe that its effects are disastrous. There are others (at least I suppose there are, but they keep very quiet about it) who believe that its effects are beneficial. I will try to examine both points of view.

The first – or earliest – objection to television is that it prevents people from reading, and the mental effort this involves. This might be called the academic objection, common among educated men fifteen or twenty years ago.

I will agree that many of the greatest intellectual and spiritual experiences cannot, and can never be, conveyed by television. *The Dialogues of* Plato, the *Divine Comedy* of Dante, the *Enneads of* Plotinus, can only be read, and the valiant efforts that have been made to read such works aloud on sound wireless have not, in my opinion, been successful. But the members of the Athenaeum who, after a good dinner, go home to read the *Timaeus* are, by any standards, a very small minority. The question is, does TV debase the intellectual currency of those who might otherwise have been using their minds? To begin with, does it really prevent them from reading? One million copies of Dr Bronowski's *Ascent of Man* have been sold. Without TV it might have sold six thousand. And is reading the only way of acquiring information or great experiences? It certainly wasn't in Antiquity, or in the Middle Ages. It is ironical that universities, which have so strongly condemned TV in the interests of reading, have clung to the mediaeval tradition of instruction by lectures, although every undergraduate knows that he would get more out of a book.

What would the mass of people have been doing in the evening before TV? They would have been engaged in the kind of dismal and repetitive conversation that we know from the plays of Mr Pinter, or from the wonderful, but horrifying programme *Till Death us do Part*. There may be – we have no means of knowing – a middle stratum who would have been reading or practising some hobby, such as building model boats, that stretched their minds and increased their skill. This hypothesis is, I think, the strongest argument against television; but it can be no more than a hypothesis.

The second objection is that television leads people to live in a fantasy world. Personally I do not find quite enough fantasy on TV programmes. I find the majority of plays all too realistic. I have led a sheltered life, and it is only in the last few years, watching television, that I have begun to realise how other people live. I think these depressing pictures of average life must be fairly correct, because no-one ever protests against them on that score. But I do agree that many of these plays, or serials, are trivial. This vulgarisation is a charge frequently made against TV by people who have not tried to find out what unevolved people read or talked about before TV existed. The age of Bible-reading was long since over, and I fear the answer is that they read the evening papers, did the pools, and occasionally studied what the stars foretold. On a long train journey one may observe that people are content to go all the way from London to Edinburgh with no other literature than a copy of the *Evening News*. Certainly they talked. They talked about their neighbours. Whether what they said was more or less trivial than what they see on TV cannot be assessed; but it will have probably been more malicious. Nominally it *will* have been concerned with real people, but very often those people will have been re-created according to an envious fantasy quite as unreal as anything on TV.

Still, the vulgarity of many popular programmes, especially those in which harmless people have been persuaded by some jackass of a compere to make fools of themselves, is really degrading – and it is no good saying, as is frequently said, that viewers are at liberty to switch them off, because they may be precisely the programmes that their children, or their relations, wish to see.

One of the accusations frequently made against TV plays and serials is that people tend to identify themselves with certain characters. This, although much pressed by psychiatrists and sociologists, is not a serious objection. We all identify ourselves with charcters in the novels we *read*. How many girls reading *Anna Karenina* or *Jane Eyre* have identified themselves completely with the heroines? It is quite difficult for any of us to find out what we are really like, and I do not claim to have done so myself, even although I have written an autobiography. I think there is no great harm in believing for a time that one is like the creation of a great novelist – or even the writer of a script – unless one identifies oneself with a criminal or gangster.

Which brings me to two of the most serious charges brought against TV; its impact on violence and on sexual licence. When I was Chairman of the ITV I was so much concerned by the effect of TV on violence that I asked for an investigation by a trained psychiatrist. The result was the report by Dr Himmelwhite which argued that TV had no effect at all on the increase of violence. I would call this a specialist aberration. I know that young males have always been aggressive; and I know that such episodes as public hangings were gleefully attended in the 18th century. But I refuse to believe that children watching TV programmes for at least eight hours a day that contain over 80 per cent violence cannot be to some degree influenced by what they see and tempted into self-identification. There are two kinds of violence. One might be called 'chivalric' violence, of which the classic example is a shoot-up in a saloon in Dodge City. This does no harm at all, any more than the accounts of Samson's exploits – 'and he went down unto Ascalon and slew thirty of them' – did any harm to our Bible-reading forefathers. The other might be called casual and sadistic violence: when in a play on contemporary life one of the characters kicks an old man to death. This is indeed an accurate picture of modern life, but I think that to represent it on TV is to make young people feel that this is an accepted, or even an admirable, way to behave. I hope the reader will agree that it is not. The urge to commit acts of terrorism and accompany protest by violence has spread all over the world, like a sort of infection, and to some extent this infection has been carried by TV.

With sexual licence I enter a far more debatable area, because there are many people who believe that one cannot have enough of it. I do not agree. I believe that in a healthy and creative society sexual intercourse should be treated as something glorious, and not treated like buying a cream lolly. Although I do not always agree with her, I regard Mrs Mary Whitehouse as one of the few people in this country with any moral courage; one of the few who has dared to oppose the intellectual establishment which has stimulated a degraded expression of the popular will. I am afraid it is a hopeless cause. A television programme without a sequence showing two people in bed together would scarcely go on the air. And of course this spectacle of licence is liberating (it is said to 'remove inhibitions') to millions of viewers. Christianity triumphed very largely because it imposed inhibitions that had vanished from the Greco-Roman world. I do not know what the biological effects of this unlimited promiscuity will be. But I do know that the effect on poetry, drama and opera will be very bad. The literature of licence – Apuleius, Petronius, Casanova, is limited and unimportant compared with the great poems in which sexual intercourse is an almost unattainable aim. From the *Roman de la Rose* to *Tristan and Isolde* what sublime struggles have preceded the ultimate union. Without the feeling that fulfilment involves the fateful defiance of some vast moral obstacle half the great poetry of the world could scarcely have been written. Perhaps the last words on the subject are still those of the poet, Burns, who had reason to know: 'But, ach, it hardens a' within, and petrifies the senses.' However, I do not think that television can be blamed for this state of affairs. It originated, like most of our present evils, in the rebellious speculations of intellectuals, who in this instance met with an enthusiastic response from commercialism.

Up to this point I have been concerned with television programmes in which people really look at the machine. But I know that in many households TV goes blaring on without anyone paying the slightest attention, and that people should become dependent on this wish-wash of nonsense as a background to their lives is often put forward as another argument against TV. As music is a form of language, the sound of stupid music depresses me as much as stupid talk. But I must allow that many writers and philosophers, including Sir Frederick Ayer himself, tell me that they can only think and write to a background of popular music. So why should not housewives do their work to a similar accompaniment? I once said to Mr Prince Littler, then the most powerful figure in show business, that TV would be more tolerable if it were not for ladies with broken hearts. 'Sir Kenneth,' he replied, very seriously, 'they are the curse of our profession.' Clearly these outpourings of sound satisfy a popular need, and, as I have said, we must accept the fact that TV is basically a popular medium. Public men spend a lot of time braying about democracy, and this is one form in which they get it. They had better learn to like it. However, I am under no such obligation, and will confess that I do not think this continual deluge of idiotic sound can do people much good. It must, to some extent, deaden their awareness of other things that are happening around them. I remember in Delphi how young visitors to that sacred place carried with them blaring transistors, and as they walked through one of the most beautiful landscapes in the world, the deep valley of olive trees below them, the shining marble theatre above them, down to the Castalian spring, they seemed occupied only with the sounds that came from the little boxes in their hands.

Having said some, although perhaps not all, of the things that can be said against television, I will now try to put the opposite point of view. In doing so I am afraid I shall have to use the words 'the average man'. I know quite well that there is no such thing: that every man and women is an individual human soul. But I must use some expression to cover the collective – and often very similar – experiences of a quantity of men and women, and 'the average man' seems to me the shortest and least objectionable. So let me say that the first thing which TV does for the average man is enormously to widen his range of knowledge of the real world. This is by no means a pleasant experience. Nearly all of it is concerned with wars, famine, riots, explosions and other forms of violence. We suddenly realise that they are taking place everywhere all the time. This is not simply due to the film makers' love of the sensational although this is a factor, but because life has always been like that. Television in Merovingian times would have been even more appalling. We must realise that we inhabit a terrible world; but it is our world, and we are part of it. The realisation could never have come to us so vividly without the visible image. And here let me express my profound admiration for the cameramen who make these events visible to us. When we see a gun battle in Vietnam or in Belfast, or a riot in Lisbon, we tend to forget that there was a man with a camera in the middle of it all, with bombs going off and bullets whizzing round him, risking his life in order to get the best possible pictures. Of course this 'one world' is not confined to horrors and disasters. I doubt if people would accept pictures of the ordinary humdrum, peaceful activities of our own country; but we are prepared to look at them if they are exotic. David Attenborough, who has done more to humanise television than anyone in England, has shown films of his beloved Indonesian Islands which reveal long established harmonious societies, that make us think as ruefully of our Western world as did Bougainville when he first discovered Tahiti. And I must add that the spectacle of what used to be called savages is not always humiliating to Western man. No doubt the behaviour of Catholics and Protestants in Northern Ireland lowers one's opinion of literate humanity. But to look at David Attenborough's series called *The Tribal Image* is to feel that, after all, we in Europe have gone some way. Without TV we should never have realised what a genuinely primitive society, with all its cruel, crazy and monotonous taboos, was like. They make our own taboos seem relatively harmless.

In the attempt to establish one world I think one of the most valuable things in recent television history was Alistair Cooke's *America*. English people are generally ignorant of American history, and prejudiced against Americans. Led gently by the hand of this great charmer, they must have had their sympathy increased. I am one of those who believe that an understanding between England and America is the essential element in a tolerable future, and I doubt if anything could have been done to promote it more artfully than Alistair Cooke's programmes.

One world. And this not only in a geographical and social sense, but in a biological sense as well. The most illuminating and educative features of English television have, I think, been what might be called 'nature films'. Their subjects range from whales to insects, from storks to elephants.

They are made with incredible expertise; various films showing *how* they were made left me marvelling at the skill and patience that men will devote to something they think worth while. These films have been enlightening, moving and sometimes frightening. Often we feel how close we are to the warm-blooded animals aspiring to the same virtues of courage, loyalty and love of our children. At other times we see how horrible nature can be, especially in the deep sea. This is de Sade's nature with a vengeance. The most blood-curdling films were not about animals, but about flowers, which trapped and killed their tiny victims with the ingenuity of the Gestapo. Everyone must have learnt something from these programmes, and they were seen by millions of people; but how much have they taken in? How much has the old, sentimental, anthropocentric, Wordsworthian concept of nature, been modified? How far does it still mean little more than a faithful dog and a fine walk on the downs? It is very hard to remove deeply rooted preconceptions, even by the vivid medium of television.

'And next to nature art'.

Well, the English, unlike the Americans, are not as keen on art as they are on nature, and it is inevitable that art programmes on English television are not very numerous or very successful, except for those concerning collecting and, by implication, bargain hunting. Nevertheless, I think that enjoyment of works of art could be – has been – increased and diffused by television far more than by books and ordinary lectures. The reason is that the camera can concentrate. It can move in and out and pick up details. In the end the critic or expositor can say very little that is helpful, but he can say 'look at this', and the camera *can* look at it.

My own experience has been that when I spoke about art on commercial television, which was then outstandingly – almost exclusively – the popular channel, ordinary people were grateful to have their eyes opened. I used to receive most extraordinary comments from taxi-men and porters (in the days when there were still porters). A friend of mine in a pub in Covent Garden thought he was dreaming when he heard the market workers discussing the merits of Caravaggio. Curiously enough, the later art films I did, although much better made, and in colour, had less effect. I suppose the novelty had worn off. But even so they were seen by many thousands of people who would never otherwise have looked at a work of art or architecture; and I believe that a new expositor, with a fresh approach, might once more open up this branch of human experience. My only fear is that, in attempting to make art more interesting, he would make his presentation too gimmicky. Production gimmicks are the curse of television. Of course directors love them, and critics praise them, but I believe that the average man wants to get on with it or to be informed. Television, as I learnt early, to my cost, is at its best when it deals with information, not ideas.

To this there has been one remarkable exception, Dr Bronowski's *Ascent of Man*. His series was not free from the gimmickry to which I have referred, and no doubt his director thought its success was due to the forging of Japanese sword blades and other picturesque superfluities. The proof that this was not so is that when the series was published as a book, without sword blades, it remained at the top of the best-selling list for almost a year. That Dr Bronowski could hold an audience spellbound while explaining the origins of Greek geometry proves that, with a grasp of one's subject, and a real love of communicating with people, ideas *can* be made digestible by the average man.

Dr Bronowski's programmes, and another series which I will not name, were attacked in intellectual circles as containing a lot of half-truths. This may be partly due to the exclusiveness of a jealous priesthood. It is irritating to see and hear things that have taken scholars much pains to discover or analyse, put in such a way that the average man can understand them. I know the feeling. I was a scholar in my youth, and I much dislike facile popularisations. On the other hand, I believe that practically everything, except certain aspects of science, philosophy and theology, can be put in such a way as to be comprehensible to a large number of people. It takes a little thought. You have to clear your mind, and translate scientific jargon into plain language. But I believe that in the present age it is the scholar's duty to do so.

The Richard Bradford bequest specifically mentioned the arts in their relation to society, so I must now consider what British television has done for the art of drama. I think its record is very impressive, and in some ways surprising. Given the voracity of television, it was inevitable that English – and European – literature would be ransacked for subjects for serials; but one would not have expected the adapters to fly higher than the Forsyte Saga. At least it seemed natural that producers would look for books with lots of punch: and in fact Zola's *Nana* was extremely effective. But the extraordinary thing is how successful they have been with novels in which the 'story line' might seem to a reader too tenuous or opaque. I am thinking particularly of the astonishingly successful adaptations of Henry James's novels. *The Portrait of a Lady* was condemned by Robert Louis Stevenson as being too low in key and undramatic. It made a most moving serial. *The Golden Bowl* was often spoken of with derision (even by Edith Wharton) as being the last word in diffuse obscurity. It was at least as effective as the *Portrait of a Lady*. What a curious irony that Henry James, who always knew that he was essentially a dramatic writer, and wasted eight years of his life writing unsuccessful plays, should have been revealed by this new medium. One must add that the success of these plays, and many other pieces of serious drama on TV, was partly due to the really marvellous performances of the leading actors and actresses. I remember when setting up the ITA many people told me that we should never find enough good actors and actresses to work for three channels. What nonsense (as I said then). In fact,

TV drama has created a new race of actors (and still more actresses), who have developed a subtle and natural technique that makes some actresses of the older school look heavily theatrical.

Television has also brought forward a school of dramatists. Through a television play, *The Tea Party*, Mr Pinter first made his name as the most interesting playwright of our time. It was seen by eight million people. That terrible and marvellous production *The Battle of Culloden* was also seen by eight million. These are figures that cannot be laughed off; and if space permitted I could add to them the works of other remarkable dramatists that were seen by almost equally large audiences.

I must now what they call in Parliament 'declare an interest', which makes me too lenient to TV entertainment. I enjoy plays about detectives and policemen. I make this confession with the less shame, because two of the chief poets of our time, Robert Lowell and Dylan Thomas, both told me how much they enjoyed detective dramas; indeed Robert Lowell was an enthusiastic admirer of Perry Mason, which is going to the limit. One could invent all kinds of psychological explanations of this weakness, as for example that men of letters, leading a sheltered life (although neither Robert Lowell nor Dylan Thomas led sheltered lives, in a conventional sense) enjoy living in the fantasy world of criminal investigation; but the real reason is that one enjoys the suspense and surprise of a good story, and this is more often achieved in a 'tec than in an ordinary play. It was a good day when *Z Cars* began to use the English tradition of documentary film for police stories; and curiously enough it allowed for the creation of real characters, which is not always achieved in a more self-consciously 'arty' play. Personally, I do not like 'tecs that try to be too deeply psychological and I am bored when they become too specialised and tend to become too complicated, as has recently happened to *Kojak*.

I enjoy television entertainment because it now gives me

the opportunity to see old movies that I had missed or partly forgotten. I have always believed that for thirty years more talent went into the making of movies than into any other form of art, and that *Citizen Kane, The Grapes of Wrath*, or many others, would survive when most of the novels and plays of the epoch were forgotten. Now, thanks to television, they can do so. As for the great early Westerns: they are the folk epics of our time. The heroes are genuinely heroic, the forces of evil banded together against them seem to be insuperable, they win through by courage and resource worthy of *El Cid*; and the whole action takes place in a beautiful white landscape that is a delight to the eye. Altogether the landscape backgrounds of television are a continual pleasure. I often wonder what the great painters of an earlier age would think of modern art. Not much. But how Turner, who used to sit for hours looking at the first photographs of the Niagara Falls, would have relished the light and colour and scenic variety of the television screen.

Finally, a word about television and public affairs. I notice that Mr Robin Day, who must know more about it than anyone, thinks that interviews and discussions are not full enough because (in a telling phrase) 'facts interfere with the story line'. Perhaps I am alone in thinking that we get our fill of interviews and discussions. I watch *Newsday* every night, *Panorama* and *This Week* and the excellent *Money Programme* every week, and witness all the interviews that take place on the *News*. That seems to me enough. The political interviews are sometimes disappointing, because politicians cannot afford to tell the truth. They are bound to think of what may be picked up and used against them. They are not addressing the viewer so much as members of their own unions or parties. This is true of all public speakers, particularly of international conferences, UNO, etc., and is in no way peculiar to television. What is peculiar to television is the discussion or debate between two or four representatives of opposite views, and although, as we are told, their answers are often curtailed, the very fact that they can put opposing views and argue them out, is surely an enormous advance in the direction of (and for once I think one may use that poor, battered word correctly) democracy. It may well be that efficient government will ultimately be achieved without popular participation; and if that were to be true TV interviews and debates would be forbidden. We are committed – for the time being at least – to an opposite principle.

As a matter of fact popular intervention in government is minimal. Decisions are made and executed almost entirely by bureaucracies which pay no attention to the popular will at all. Still, the fact remains that people can voice their discontent, and this is perhaps some kind of brake on the bureaucratic machine.

I believe the best thing about public affairs broadcasting is that people can form some sort of impression of the men who nominally control their destines. This is not quite true, because some people feel more at home with the medium than others. Still, I do think that from TV appearances one does get a feeling of what a man or woman is like, far more than one would have got from speeches on a platform. Very few out-and-out scoundrels make the impression of honest men on television.

It will by now have become apparent that I am a supporter of television. I recognise its shortcomings. For a majority of people it is, half the time, a trivial opiate. It can never become an elitist medium, and I doubt if any great art can ever flourish without some kind of an elite to absorb and direct talent, all the more so now that talent is drawn from a so much larger catchment area. It can do little to give us spiritual comfort and enlightenment, and Heaven knows we are in need of both. In other words it cannot be a substitute for religion, as love of nature was in the early 19th century, and love of art was, to some extent, in the late 19th century, and so-called 'ideologies' have become today.

But in all these ways it is not the cause, but the illustration, or visible projection, of a spiritual and intellectual decline which has overtaken us in the last thirty years. Whether this is due to the effect of two vast wars, or to diversion of all the best brains on technology, is a problem for future historians.

Given these conditions, I think that the amount of good that British television has done is remarkable. Far from having debased popular taste, I think it has always been a little bit ahead of it. It has enormously widened men's horizons; it has increased their knowledge of the world and of nature, and even whetted their appetites for art and ideas. It has produced works of dramatic art and familiarised people with great works of literature, which they never would have read for themselves, and have often done so after seeing the TV adaptations. It has alleviated solitude; relieved boredom; and I hope I may be allowed to say, even in a context so serious as the Richard Bradford Lecture, it has made us laugh. The best episodes of *Dad's Army* were as funny as the play scene in *A Midsummer Night's Dream*. Surely that is something for which we may be grateful. But perhaps I should not end on such a frivolous note. Even if TV be three-quarters a symptom of public feeling, and one quarter a cause, that quarter is important enough to require continual scrutiny. It could become a real menace. Both political parties hate it. If anything disagreeable happens they both blame what they call the media. The Tories think, with some justification, that it is subversive, and helps to promote the class war. Labour dislikes it because it is critical and out of control. That such an important element in society should be 'out of control' in a time when practically every department of life is controlled, is very troublesome; and successive committees of enquiry and commissions have tried to see how it could be trimmed in the interest of bureaucracy – so far without success. If ever their efforts do succeed, and we have a state controlled television, I fancy we shall find ourselves regretting our old vulgar higgledy-piggledy, from which, after all, so many good things have emerged.

24th October, 1975

LECTURE II

That Famous Greek 'Wholeness'

H.D.F. Kitto

My commission is to say something about one of the major Greek arts, the art of tragic drama, and to do that not from the aesthetic or literary points of view but as a means of exploring the ways, or some of the ways, in which the Greeks thought about some of the problems of life. We shall find them different from our own.

The word 'wholeness' I owe to Matthew Arnold, to his often-quoted verse about Sophocles:

> He saw life steadily and saw it whole.

Very true of Sophocles, as I hope to show, but true also, I think, of Greek thought in general, at least in what I shall call their Early – Classical period – that is, down to about the fifth century B.C. I shall be concerned almost entirely with Sophocles, but my observations would apply just as well to Aeschylus and Euripides, with slight modifications, and, with rather more modifications, to the historian, Thucydides, too.

Let me point out at once: I am considering Greek thought, not Greek behaviour. The political and social history of the Greeks was as tumultuous and fierce as that of most races; my concern here is with their deeper reflections about the human scene.

I had another reason for choosing the term 'wholeness'. We today are worried – from time to time – by the increasing fragmentation of life, by the increasing difficulty of seeing it as a whole. The reasons for that are plain enough, politically; for example it is easier to see things as a whole when the population of your State is, say, a hundred thousand and not a hundred million or more. I am not going to discuss the many reasons for that fragmentation; I wish only to assure the reader that I have not forgotten them.

But, on the subject of fragmentation, in recent years we have heard a lot, or at least quite enough, of the Two Cultures. Between the Arts and the Sciences, we are told, lies a deep gulf. I have quoted Matthew Arnold; now I am going to quote a very different writer, A.N. Whitehead; most readers will agree that anything said by Whitehead deserves to be taken seriously. In one of his books he wrote: 'Insofar as modern scientific thinking has Greek ancestry, its Greek ancestors are not the Greek philosophers but the Greek tragic poets'. Surely a remarkable statement, coming from so distinguished a source. The implied disparagement of the philosophers one can easily understand: presumably Whitehead was thinking of men like Democritus, with his atomic theory; men who made brilliant 'guesses at truth', but had neither the means nor, it may be, the inclination, to test them by experiment (though we should not be too absolute about that; some of the Greek medical writers had a very firm grip on the necessity of experiment). Whitehead knew, of course, that in what I may call the observational sciences – biology, medicine, astronomy – the Greeks were unique in the ancient world, and I need say nothing here about Greek mathematics. In astronomy, the Chaldees and other peoples of Mesopotamia had for centuries made and recorded accurate observations, but for practical reasons – agricultural and astrological. The Egyptians knew and exploited the mathematical fact that $3^2 + 4^2 = 5^2$; it was the Greeks who, by no means despising the practical application of such knowledge, wanted to know Why; with the result that Apollonius of Perga four or five centuries later wasted his time writing a treatise on Conic Sections. All that, and much more, will be familiar to the reader, and certainly was to Whitehead. But what is arresting in his dictum, for me at least, is his mention of the tragic poets. Does it make sense?

These two quotations, then, serve as my starting-point: that Sophocles (and, by extension, the other two surviving tragic poets) saw life steadily and saw it whole; and that, in a tentative way, they can be regarded as distant precursors of modern scientific thinking.

Before I really get going I must say something that will be relevant to what follows. Some students of Greek Religion and religious thought have said: we can trust inscriptions, we can trust what we know about religious cults, but we cannot trust the likes of Aeschylus or Sophocles, because obviously these were quite exceptional men, no sure guide to what the ordinary Greek thought and believed. That sounds sensible, but isn't. For the dramatic Festival of Dionysus, in Athens, was one of the great events of the year, the only occasion in the year when tragic plays were presented. It was the duty of the relevant Minister of State (as we might say, except that he was directly responsible to the assembly) to supervise the Festival, and in particular to choose, out of all aspiring poets, the three who should have the honour of competing – as we might say, poets-laureate for the year. Further, we should bear in mind that the theatre was built to hold fifteen thousand spectators, or thereabouts, and that when the total *citizen* population was about twenty-five thousand. Others could attend, of course, but still, it is substantially true to say that the poets were selected citizens presenting their work, on a great annual occasion, to their fellow-citizens; not just brilliant individuals performing to an *élite*.

In these circumstances we would not expect the dramatists to deal with trivial or eccentric themes, nor to propound ideas which were above the heads of their large audiences – still less offensive to them. No doubt there were dull plays; three tragedies per annum, from each of three poets; nine hundred over the whole century. We possess thirty-one of them, and we have good reason to think that these are of the best, but no reason to think that all were of this quality. But we do know that Aeschylus and Sophocles, less so Euripides, were the most admired poets of their time. That, too, is proof enough that the ideas they expressed were not out of the reach of their audiences though they may, at first sight, be out of our reach. Twenty-four centuries can bring many changes.

It was a religious festival; it opened with a procession and sacrifices in honour of the presiding god, Dionysus. Further, a god or gods always have some part in the action of the plays, whether visibly, on the stage, or by constant implication, from behind or above it. Therefore, reasonably, we call it 'religious drama'. That has a comforting sound, but how 'religious' was it? This is where our troubles begin, and they have been many and vexing. For I must not conceal the fact that since the serious study of the plays began in modern times, say, two centuries ago, scholars and other readers have been at sixes and sevens over the interpretation of many of them, often flatly contradicting each other. One fairly common refuge has been the assumption that the dramatists were, in spite of their other high qualities, not very good at designing their plays; they were apt to fall into elementary blunders of construction. But, since an outstanding quality in the other Greek arts is a sure control of form, that explanation is not very convincing. Much more likely that we, belonging to a different age, are confidently expecting the wrong thing, and then blame the poet when we cannot find it. It may well be that those twenty-four centuries have deposited on the Greek stage certain litter which gets in our way. If we can distinguish and remove that litter, using the assumption that the dramatists really did know their job and were as good craftsmen as the other Greek artists, then perhaps we shall be able to come a little closer to their minds; and if we find that the removal of the litter makes the plays more elegant and exciting, our confidence will be all the greater.

Very well; what about that term 'religious drama'? Sometimes it leave us quite happy. In the *Antigone*, for example,

Sophocles writes for his chorus a stanza which may be rendered like this:

> Thy power, Zeus, is almighty, no
> Mortal insolence can oppose Thee.
> Sleep, which conquers all else, cannot overcome Thee,
> Nor can the never-wearied
> Years, but throughout Time
> Thou are strong and ageless,
> On thy own Olympus
> Ruling in radiant splendour.

Here is religious poetry of exceptional quality. With a few verbal changes it might have come from a medieval Christian hymn.

We feel equally at home when we read, in the *Agamemnon* of Aeschylus, another such invocation:

> Zeus, whoever Thou art, if this name please Thee,
> By this name I invoke Thee.
> As I ponder all things, my mind
> Can form no picture but of Zeus,
> If it would throw off the vain burden of thought.

And later in the same play the chorus invokes Zeus as 'the Cause of all things, the Worker of all things'.

That is fine, but it is not long before we begin to wonder, because with some frequency all three of these religious poets make their gods do some very ungodlike things. In consequence there is some disposition to say that the religious ideas of Aeschylus were still backward, that Sophocles was an artist and therefore didn't think much (which is more or less what Plato said, too) and in any case was a bit complacent over his religion, and that Euripides was a sceptic and revolutionary. But the explanation is far simpler than that – or so I shall argue: it is merely a matter of translation. I cannot say that our word 'religion' is a misleading translation of the Greek word, for the reason that the classical Greek language had no such word at all. Surprising, but true, and I shall return to it later.

Another is our word 'god' as a translation of the Greek word *theos*. We have to use it; our language offers no other, except some clumsy periphrasis which would sound like a footnote. This really is a difficult case: sometimes 'god' is a perfectly good translation, as, for example, in the passages I have just quoted from the *Antigone* and *Agamemnon*, but quite often it puts us on the wrong road altogether, as I will try to show, at intolerable length. If we stick to the word 'god', we shall stand no chance of making sense of Whitehead's dictum. Yet another such word is the Greek *dike*. Our habitual translation is 'justice'. Sometimes that, too, works very well, at other times, no. What do we do when we find one of the early Greek philosophers, writing about the nature of the physical cosmos, saying, 'Things are for ever taking retribution for each other's injustices'? For us, justice and injustice are ethical concepts, but obviously *dike* and its opposite, *adikia*, have no ethical connotation here. Anaximander was indeed a philosopher, but he was not a fool as well. A more sensible translation would be: 'Things are always making restitution to each other for encroachments'. In winter the nights are longer than the days, but the days get their own back in summer. In the end, things even out. This entirely non-moral use of *dike* will meet us when we come to consider one or two of the plays; our word 'justice' may send us far astray.

After these preliminaries, let us look at a specific play. The snare hidden in the word 'god' for *theos* will become apparent as we go along. For several reasons I have chosen the *Electra* of Sophocles, the chief one being that it is a first-rate play, regularly undervalued, because misunderstood, by modern readers, including Greek scholars. Briefly, the broad context is as follows. Agamemnon had fought and won the Trojan War, to do which he had to sacrifice to Artemis his daughter, Iphigeneia. (I would say a lot about *that* if space permitted – as I know it does not.)

In revenge for that, his formidable wife, in concert with her lover, Aegisthus, murdered him. His infant son was smuggled abroad by his sister, Electra, lest they should kill him too. When the play opens, Orestes, the said son, has reached manhood and has resolved that he must go back to Mycenae and avenge the crime: to kill the guilty couple, recover his patrimony, and deliver his lawful subjects from a lawless and oppressive tyranny. This he tells us in the first speech; also that he went to Delphi to ask Apollo how he should set about it. (Since some eminent scholars have assured us that he was commanded by Apollo to do it, it is just as well to read the text again and discover that he wasn't: the idea was his own.) Apollo concurred and gave him the advice for which he had asked: 'Do it not by open force but by stealth, and give them what they have earned.'

(The word that I have translated as 'stealth' covers anything from legitimate stratagem to downright treachery. It is *dolos*.) So here he is, in Mycenae, accompanied by an old slave who has been with him since infancy, and a faithful but silent friend, Pylades. He outlines the stratagem which he has devised. The old man is to go to the palace and tell them that he, Orestes, is dead, killed in a chariot-race at Delphi. That will throw the two criminals off their guard. Then Orestes will present himself at the palace, pretending to be an obliging stranger who is bringing home the ashes of Orestes in a proper funerary urn. They will certainly let him in, and he, with luck, will do what has to be done.

The three men leave the stage. Electra comes out of the palace, singing or chanting a monody of pure grief for her father, mixed with prayers for vengeance and for the coming of Orestes to execute it. From this moment until the very end of the play it is Electra that dominates the stage. We see her in sharply-contrasted scenes: first with the chorus – sympathetic women of Mycenae – then with her sister, scorned by Electra because she is content to live at peace with her father's murderers. Then a scarifying scene between Electra and her mother; after which the old slave arrives. He plays his part superbly. He tells them his false tale of the chariot-race brilliantly and convincingly. Clytemnestra is triumphant: the threat of vengeance is removed; now she can live in peace. Electra, naturally, is shattered; what she has been living for all these years is now impossible. But after a short time she recovers; in a positive ecstasy of determination she announces she will attempt the deed herself: *she* will kill Aegisthus, if she can. If she fails and is put to death – well, death is better than life, now. But Orestes arrives, with his silent Pylades and apparently two or three other attendants. Naturally, brother and sister do not recognise each other; they have not met since he was an infant. (Clytemnestra has of course returned into the palace with the very welcome old slave.) After some talk Electra begs Orestes to let her hold the urn. He gives it to her, and over it she makes a most moving speech, the theme of which is that all the love and devotion that she had lavished on him when he was a baby, all the hopes she had that some day he would come back to right the wrong, have been reduced to nothing. It was all in vain. That of course leads to the recognition and to a long outburst of joy from Electra. At last Orestes and Pylades enter the palace, to deal with Clytemnestra. Aegisthus, we have been told, is in the countryside. Electra follows them for about half a minute, then emerges, and stands where she can see what is happening inside, and so tell the chorus. It is now that we fully realise the terrible depth of her hatred. We hear Clytemnestra scream once; then again: 'Oh, a second blow!' Electra cries, 'Strike her again, if you have strength enough!' Very soon Aegisthus arrives; he had heard that Orestes is dead. Like Clytemnestra earlier, he is triumphant: the only possible avenger of Agamemnon is dead. Now, as tyrant, he is secure; now, even the irreconcilable Electra must submit. 'Is the news true?' he asks. 'Yes,' says Electra, 'the messenger brought not only the news, but the body too.' 'What, is the very body to be seen?' – 'Yes.' – 'Let it be brought out.' The body is brought out, decently covered in a drapery. 'I too must mourn,' says Aegisthus, 'after all, he was a kinsman'. So he turns back the facecloth, and finds himself gazing at the dead face of his wife.

He sees at once that he has been trapped, that his hour has come. 'You,' he says to the stranger, 'you are Orestes!' 'I am.' 'Then let me speak.' But Electra intervenes. It is her last utterance in the play. 'No, no spech from him! Kill him at once! And when you have done it, throw his body out, for dogs to devour.' It is horrifying. The end is swift. With a grim show of courtesy, Orestes drives him into the palace, to die, as he says, 'on the very spot where you killed my father.' Nothing remains but a brief comment from the chorus, as it retires. Three verses only, and to this effect: 'Well, you have achieved it at last, and a good job too'.

And what do we say about it all? Superficially, an exciting revenge-play, with a terrific climax – though is it really a tragic climax, or only theatrical fireworks? There is also the vivid and sustained character-study of Electra, but is it not a little disconcerting that a portrait of a tragic heroine should end as harshly as this one does? Its modern critics have been sorely puzzled by the play; some hostile. The distinguished American scholar, who is also a poet, Richmond Lattimore, has said of it, quite correctly, 'The play has left its various critics to come to the most various conclusions, including the conclusion that there is no conclusion'. One or two have called it an 'enigma'. But surely it is not in the least likely that this elegant and powerful dramatist should have presented his vast concourse of fellow-citizens with an enigma? It may be an enigma to us, but that is another matter.

It is for that reason that I chose this particular play for discussion here. If we find it an 'enigma', may not the reason be that there is in it something so distinctively Hellenic, something so foreign to our way of thinking that we do not grasp it? What are the obstacles? Two in particular.

One is that the play drives forward to its consummation, which is the awful act of matricide – for we need not worry too much about the killing of Aegisthus. That, as many a commentator has said, involves a grave moral problem, and Sophocles does not give it the slighest attention. Some forty years earlier Aeschylus had dramatised the same story, but he did face the moral problem. For him, the matricide was indeed a necessary act, but it was one to be purged although in the end Orestes was absolved; but Sophocles ends his play with it, and, far from hinting at punishment or purgation to come, he winds up with that flat comment from the chorus: 'Good! O.K.' How could Sophocles be so insensitive?

Our second stumbling block is the role of the gods in the play. We happen to know that Sophocles, personally, was a man of piety. How, then, could he have represented the gods, any of them, countenancing matricide? Some critics have tried to evade that difficulty by trying to push them out. Thus, one has recently written: 'Sophocles takes us at a run past Apollo'. He certainly does nothing of the kind. It will be enough if I mention two conspicuous incidents in the plot. Early on, Clytemnestra, we are told, had had an alarming dream that seems to portend the restoration of the Royal Line. She is badly frightened, and comes to offer solemn sacrifice and prayer to Apollo. The prayer is blasphemous: she prays that Apollo shall avert anything of evil fortune in the dream, turn it against her enemies, safeguard her in the continued enjoyment of the wealth and royal power that she now enjoys, and (without putting this into words) grant that her son may never live to come back and take vengeance upon her. As soon as she has ended that impious petition, the old slave, with apparent casualness, enters and gives his false report of Orestes' death. Now, if Sophocles did not expect his audience to take that as Apollo's immediate answer to the prayer, then as the saying goes – I'm a Dutchman, and you well know that I am not.

Then as Orestes and Pylades enter the palace, Electra makes an impromptu sacrifice to the same god at the same altar. She prays that he will further their present purposes, 'and show to all mankind what chastisement the gods inflict on those who practise wickedness'. And that prayer too seems to be answered immediately. Surely Sophocles is not simply playing at theatre? But if not, how do we answer that urgent question of how Sophocles could possibly represent the gods as favouring this act of matricide?

This seems to be the moment to take up the thread which I left dangling a little while ago: the meaning of the Greek word *theos*. As I said then, our translation 'god' is sometimes good enough, but sometimes quite misleading. *Theos* covers a wider range of meaning than 'god'. Let us abandon, for the moment, the *Electra* and try to expound the meaning of *theos*.

Here, for a start, is a pretty little problem, Homer, in *Iliad* IX, is describing how Achilles prepared supper for three friends. The dish was what is now called souvlakia or shish-kebab. He has cut up the meat, impaled it on the spits, puts it on the fire and sprinkles it with 'divine' salt, and I am not playing tricks in giving that translation: Homer's adjective is *theios*, the adjective derived immediately from the noun *theos*. When I was at school the commentary that we used – quite a reputable one – explained that salt is 'holy' for two reasons: it purifies, and it was used in sacrifices to the gods. That shows how easy it is to stumble over the aforementioned word. For one thing, salt preserves but does not purify; and for another, this was no sacrifice; simply a friendly social occasion. Achilles, as cook, was naturally using ordinary table salt, plain sodium chloride. But what makes sodium chloride 'divine', or 'god-like'? Was Homer being silly, or are we mistranslating him?

Let me tease a little further. The earliest Greek philosopher that we know of – Thales – asserted, so Aristotle tells us, that the world is full of *theoi*. Thales was certainly not being religious; he was doing his best to be what we used to call a Natural Philosopher. His idea seems to have been that the world, the physical world, was not composed of inert matter operated by some external power but had within it its own life, forces, laws. Such *theoi*, clearly, have nothing to do with religion.

We might also consider this. Our own Creation myth tells us that, 'In the beginning, God created Heaven and earth'; the Greek myth was characteristically different, that in the beginning Earth created the Sky, the gods, and mankind, and Earth, personified as Gaia (the same word), remained the oldest of the *Theoi*, the mother of all.

I am following here the cosmogonical poem of Hesiod, *Theogony*, or the *Generations of the Gods*. In the poem, all the great natural phenomena, Day and Night, Mountains and the Sea, were born of the Earth, and therefore themselves *theoi*, or at least *theioi* 'divine': and not only physical phenomena like those, but also many others that are permanent or recurrent facts in human experience: Old Age, Anger, Hope – anything, we may say, which is not created or willed by man, anything that seems to belong to the very nature of things, good or bad, nice or nasty. Anger will serve as an example. People are always getting angry; and presumably always will. We do not choose to get angry, we merely find that we *are* angry. Anger is omnipresent, always recognisably the same phenomenon, therefore *theios*. To be sure, we can do nothing about Old Age, but we can keep anger at bay. Such things are part of the unalterable framework of existence; our business is to accommodate ourselves to them, so far as we can.

And that brings us back to sodium chloride. Salt, like air and a few other things – Light and Sleep, for instance – is something without which we cannot exist. But we cannot make it; we can only find it, in the sea or under the soil. Therefore it belonged to the *theia*. We would do more justice both to Homer and to ourselves if we said not 'divine' but 'elemental' salt.

It is our own religious inheritance that makes this difficult for us to assimiliate, to say nothing of the superb figures of gods in Greek religious art. For us, the distinction between the sacred and the secular is so ingrained that we find it difficult to imagine a 'religion' that extended without a conscious break from spirituality to common salt. Yet this was instinctive to the Greek. Let us come back to Tragedy, and take a few samples which are puzzling, even unintelligible, unless we understand the *theoi* in some such way as this.

The *Antigone* is a fairly well-known play. Towards the end the prophet comes in to tell the King, Creon, that what he has done has so angered the *theoi* that they are going to pay him back in his own coin: already their Erinyes, Divine Avengers, 'Furies' as we call them – to our occasional confusion – are lying in wait for him. He has refused to give burial to the body of Polyneices, but has thrown it out to be eaten by the dogs and birds because he was a traitor. Antigone, who did bury her brother's body in defiance of Creon's decree, he has walled up in a cavern to die of starvation. In return for those actions, the gods are going to make him pay with the life of one of his own kin. It falls out as predicted. His son, Haemon, betrothed to Antigone, breaks into the cave hoping to rescue Antigone, but she, in despair, has hanged herself. Next, Creon, terrified by the prophecy, comes in to release her. Haemon, in a mad rage, tries to kill him. He fails, and kills himself alongside his dead lover. Then the Queen: she has lost one son already, thanks to Creon. She can take no more of it: she stabs herself, at an altar, and dies, cursing Creon. He is left to live amid the desolate ruins of his own house.

Now, all this is so motivated by Sophocles as to convince us that it is just what would have happened, in the circumstances; a natural course of events. The gods, it seems, have done nothing at all. Why then did Sophocles bring them in? Why indeed, if not to suggest that what is done by the *theoi* and the natural course of events are the same thing? In other words, what we have been watching is not merely what a newspaper might call 'Sensational Triple suicides at Thebes'; it *is* that, if you like, but it is also a great deal more: it relates the particular events to something permanent in human experience. To use Aristotle's phrase, it is not only what did happen; it is also what would happen.' The 'gods' have not *caused* the event. Their presence reveals something universal in it: Creon's actions have contravened something fundamental in human affairs.

Another example. In the middle play of Aeschylus' Orestes trilogy Orestes is telling Electra of the awful command that Apollo has given him – that he must kill his mother and her accomplice. He relates in frightful details that intolerable pains and penalties will befall him if he shirks it. So there you are: a god compels a man to do something; the man is a mere puppet in the hands of an omnipotent god. Yes, but wait. When Orestes has spent some twenty verses on that, Aeschylus gives him eight or nine more: in these he goes on to say that he would have to do it anyhow, even if Apollo has not spoken; and he gives his reasons; personal honour (his duty to his father), his duty to his own citizens, now suffering under the heel of an abominable tyrant, and his own financial embarrassment, for he is in exile, living on the charity of others. So that the 'supernatural' and the natural come to the same.

This leads us to a feature of the plays that is strange to us but obviously understandable to the Athenian audiences. Again and again the same action is presented on two planes at once, the human and what we call the divine. Two examples of this, also from the *Oresteia*.

At the beginning of the *Agamemnon* we are told, explicitly, that Zeus had sent Agamemnon against Troy to wage bloody war on Paris in retribution – the word is *dike* – for his crime in carrying off Helen, the wife of Menelaus, his host. So that Agamemnon too is something of a puppet? No: he is unconscious of being an emissary of Zeus; for him, and for his indignant citizens too, the war was entirely his own idea: when he is back home again the loyal chorus-leader says to him: 'I tell you straight: when you started this war, for the sake of a woman like Helen, I thought you must be nearly out of your mind.' And of course Agamemnon has to pay, with his own life, for all the slaughter that he has caused. Yet, on the other plane, it was no less a personage than Zeus himself who had started it. A strange business, surely, especially if we recall those two passages from the same play about Zeus which I quoted earlier.

Then, later in the *Agamemnon*, there is the matter of Cassandra, the infinitely tragic Trojan princess. Aeschylus gives two entirely different accounts of her death. On the one plane, we have this. When the booty from Troy was being shared out, the army gave to Agamemnon Cassandra, the very flower of the spoil. He has brought her home with him, and is so hopeful as to tell his wife to take her into the palace and treat her nicely. Later in the play she tells the chorus with what glee she had killed her husband's lovely mistress on top of him. That does not astonish us; it is just what this woman would do. But meanwhile we have been given quite a different story by Cassandra herself, and this one is made just as convincing as the other. Apollo wanted her, bribed her. At first she had said Yes, but when it came to the point she said No. Apollo was furious, and now, to satisfy his fury, he has brought her to this palace, reeking with the blood of previous crimes, to be done to death by that murderess and adultress. But I am really discussing here Sophocles' *Electra*, so that I must go easy on the *Oresteia*, and must content myself with this.

Aeschylus, in this trilogy, is concerned with *dike*, that is to say, here, with the problem of crime and punishment. Throughout the *Agamemnon*, he gives instance after instance in which a crime, or an offence, is answered by instant and violent retaliation. The crime of Paris was avenged by war and slaughter – and so on, all the way through. The end of the trilogy is going to be the establishment of a more civilised and stable way of dealing with offences – for the present one leads to unending violence and ultimately to sheer chaos. The most ungodlike behaviour shown on the 'divine' plane is there for the same reason as before: to universalise the individual acts of violence. We are, so to speak, in the presence of violence itself. We are not to concentrate on the character, motives, psychology of Agamemnon, Clytemnestra, Aegisthus themselves, except in the second degree. But that is precisely what we moderns insist on doing, in our enchantment with Tragic Heroes and the like; consequently we find the *theoi* an unintelligible nuisance. But the function of the human characters is to exhibit the violence in sharp, and as it were local terms; the function of the *theoi* is to add a further dimension, to universalise the particular. Religion has little to do with it. I will risk my neck by suggesting that what we find time after time in these plays smacks more of the scientific than of the religious mind. Is that, I wonder, what Whitehead was thinking of?

And now, if we have cleared away some of the litter deposited by the intervening centuries, we will return to the 'enigmatic' *Electra* with tremulous hope.

There are many complaints often levelled against it: gods countenancing matricide, the total indifference to the grave moral problem, the unsatisfying 'O.K.' conclusion, the fact that the whole play consists of an unedifying bit of trickery, the stratagem. But now we will look again. What overshadows the whole action is the past triple crime: murder, adultery, confiscation. We are shown how that has affected Orestes. In more detail, we are shown how it is affecting Electra. In her first long passionate and bitter speech to the chorus she describes in detail the hateful and humiliating nature of her present life, living in the constant presence of her father's murderers. It is most vivid: it reveals much of her character. Of course; if the characters are not made vivid and alive, they would not be convincing, and neither would the basic theme of the play carry conviction; but if we lose ourselves in contemplation of the characters we shall only imperfectly understand what the dramatist is really thinking about.

As I ventured to suggest earlier, one aspect of *dike* may be described as the back-swing of the pendulum. That is particularly true of Sophocles' conception of *dike*, and especially perhaps of this play. What the criminals had done generates its own recoil. That is the way of things; it is the way of the *theoi*. The recoil may be nice or nasty; that is irrelevant: it will depend on circumstances. Of course Sophocles ignores the moral problem; he is thinking of something else. Let me call attention to a few details in the play that corroborate. First, the *dolos*, the treachery, stealth, guile. The word, with one or two synonyms, occurs about a dozen times in the play. First on the lips of Apollo: do it by *dolos*; then, repeatedly, of the way in which the murderers had slain Agamemnon at a banquet; then again of Orestes' stratagem. You see? the biter bit, and in the same way. Further, you may remember how Orestes, with some little circumstance, drove Aegisthus into the palace to die on the very spot where he had killed Agamemnon. Finally, a lyrical comment given to the chorus. I have not yet mentioned it. After Clytemnestra had been killed, and Aegisthus' turn is coming, the chorus sings: 'The tide is turning: the dead are active. Those who were killed of old are now drinking the blood of those who killed them.' The ebb is followed by the flow. One might not take any particular notice of the simile but for the fact that something of the kind appears in no less than five of the extant seven plays: the idea of a pattern, or rhythm, in human affairs. In the *Ajax* summer succeeds to winter and winter to summer: in the *Trachinian Women*: in human life joy follows sorrow and sorrow follows joy, as the Great Bear circles for ever around the Pole Star. Such similies are common in his plays, and I think, significant. Sophocles, like the other two tragic poets, always had in mind the very nature of things.

He *did* see life steadily, and he did see it whole, and if that escapes us as we read him, the main reason is that we no longer expect tragic drama to have that extra dimension, but to concern itself entirely with individuals. That is the reason, incidentally, why criticism of *Hamlet* during the last two hundred years, has concerned itself almost entirely, and not very fruitfully, on the fascinating though enigmatic character of the Prince of Denmark. Some of you, if you are now stricken in years, may remember Olivier's film-version of that play, and its title-page: 'Hamlet, or the Prince of Denmark. The tragedy of a man who could not make up his mind'. I told a young American colleague of this, once, and remarked that during the action no fewer than eight of the characters die. His dead-pan answer was, I thought, perfect: 'Hm! Shows the importance of making up your mind, doesn't it?' And so with the *Electra*. If we wanted to supply it with a sub-title, we might borrow one from Shakespeare: *Measure for Measure*. Or even from Holy Writ: 'With what measure ye mete, it shall be measured to you again.'

27th May, 1976

LECTURE III

By Their Arts You Shall Know Them

Jacquetta Hawkes

I am proud – and this is not just an opening formula – to have been invited to give the third of the Richard Bradford Trust lectures. I cannot possibly equal the authority and skill of Lord Clark, or the beautifully precise subtlety of Professor Kitto. However, I can find some reassurance in the fact that the founder of the Trust first approached me because he had discovered ideas in my books that chimed in with his own.

I have another quite personal qualification for sharing in the Trust's investigation of an affinity between artistic and scientific creativity. I am the daughter of Gowland Hopkins, a founding father of modern biochemistry – and he was a cousin of the poet Gerard Manley Hopkins. Many years ago I gave a radio talk comparing the two men and finding similarities in the way their minds and imaginations appeared to work.

In those days the philosophy of science was less concerned with theories of the advance from hypothesis to verifications by experiment and more with the old Baconian belief in purely inductive methods. It was therefore more interesting to record how my father's young colleagues used to say, 'It isn't fair. Hopi always starts with the right answer and only has to do experiments to prove it.' In order to give a suitably elusive hint at the Bradfordian idea of an intuitive/observational basis for the work of both scientist and poet I will repeat a little story from my father's youth. One day he had gone to Kew Gardens hoping to find a Large Red Underwing moth. He failed, but as he walked back to the station he became convinced that he had in fact seen one, made almost invisible by its protective colouring against the bark of a tree. With his inner eye he saw this grey, fissured shape on its grey, fissured background so clearly that he retraced his steps and found that the big moth was indeed there, a beauty of the night waiting to take wing. My father told me this true tale as a parable of how an idea might come to him, emerging into consciousness from the mists of the subconscious, where it must have formed from half-aware observations, the ground of intuition.

I believe that in this emergence of an idea that will then be the nucleus of creative work, the highest kind of imaginative scientist and the artist may share a common experience. Yet my own conviction is that even at this moment of genesis there is a difference between the mental process of scientists and artists. The distinction surely is that the idea of the artist, and above all of the musician and poet, is given its magical vitality and power to rouse a response in all men, by bringing up into consciousness material from a deeper level of the psyche – far below thought and below that subliminal observation of the moth parable.

In children, among unlettered peoples, in all of us when the unconscious is stirred by deep emotion, and lastingly in great artists, the dromos leading up from the depths is more open. So Wordsworth saw the earth 'apparelled in celestial light, the glory and the freshness of a dream', while the partial closing of the passage made the vision 'fade into the light of common day.'

Although we must not forget this vital element in art that is present only faintly in scientific inspiration, there is no doubt that a parallelism between the methods of art and science, and even the Bradfordian concept of their verification, become more acceptable after that greatest revolution in human thinking achieved by the Greeks. The new detachment attained by the Ionian 'natural philosophers', the ability to look at man outside the mythological dream of ancient religions, not only initiated science but vastly increased the conscious, thoughtful, aspects of art. It also tended to make them a vehicle, though I must insist only a vehicle, for conveying social and personal moralities.

No doubt future lectures in this series will most often be devoted to this later, post-Homeric and post-Ionian half of the history of civilisation. (Note, in passing, that the 'Greek miracle' does in fact come almost exactly half way between the dawn of civilisation and today, with some 2700 years on either side.)

For my own purpose I want to say something about the visual arts of civilisations that flowered either before the intellectual revolution or far beyond its influence on the other side of the Atlantic. I shall first be concerned with sculpture and painting from Sumeria (Figures 1 and 2) from the Egypt of the Old Kingdom and the Amarna period (Figures 3, 4 and 5) and from Minoan Crete (Figures 6 and 7). Then, leaping forward two to three thousand years, I shall consider some works from pre-Columbian Mexico (Figures 8, 9 and 10).

I hope that in the long view it may be helpful to the Trust's investigations to recall these pre-scientific civilisations whose peoples lavished so large a share of their energies on arts that were completely enmeshed in their religious mythologies. To do so may make it easier to disentangle that part of artistic creation, so much more 'primitive' both psychologically and historically, which I believe to be eternally distinct from scientific method.

In a shorter view, the main object of what will inevitably be a superficial scanning of my chosen civilisations will be to see how far their visual arts express something of the nature and behaviour patterns of the peoples concerned. At the same time looking at these individual cultures, as different as lupins, delphiniums and lilies blooming side by side in a garden bed, I shall suggest some conditions that may have affected, though they certainly did not determine, this marvellous variety.

In treating civilisations as distinct entities I shall be following Professor Frankfort's view of each having an elusive yet unmistakable character that (to quote) 'shapes its political and judicial institutions, its art as well as its literature, its religion as well as its morals.' I shall also follow him in recognising that this 'form' tended to emerge complete in all its parts during a surprisingly short time: in fact that one can reasonably speak of the birth of each civilisation rather than of a gradual evolution. After this birth and what was usually a rapid growth to a brilliant maturity, the 'form' would endure for a longer or shorter time, often on a slowly declining plateau, then, probably after some violent historical blow, collapse. I think it can properly be said that all the cultures we shall be inspecting died – even though they left some seeds behind them.

The concept of the birth of civilisations is nowhere better exemplified than in those of the Sumerians and Egyptians. Despite recent discoveries of amazingly precocious Neolithic settlements in Anatolia and of early Bronze Age towns in Iran, it remains true that the first literate, high civilisations on earth were created in the valleys of the Tigris–Euphrates and the Nile. Both sprang from originally undistinguished prehistoric societies during the second half of the fourth millennium B.C. The Sumerians led the way in true urban living, in temple architecture and in the invention of writing. The Eygptians, having been stimulated by strong influences from Sumeria, caught up with – and in some ways surpassed – the Asians as soon as their land had been unified under their first dynasty of kings. By 3000 B.C. both civilisations had left the nursery and assumed their own personalities.

From our point of view it is most satisfactory that these two contemporaries, having much in common in their valley settings, in their economic base of irrigation agriculture, in kingship and their belief that all human life was dependent upon divine powers, still produced what were in many ways contrasting 'forms' of civilisation. These contrasts are delightfully manifest in their visual arts. After discussing a small selection from their creations, I will suggest the more or less mundane factors that can partly account for the differences – while leaving much to the blessed and still mysterious freedom of the human spirit.

Sumerian art has a unique interest in the earliest expression of the vision and needs of a people achieving high civilisation. The sculpture in stone and bronze, the cut-outs in bone and shell, the cylinder seals of the late fourth and first half of the third millenia accompanied the emergence of priestly and royal establishments with their temples and palaces, and their gradual development of writing. In this theocratic society inevitably almost all works were created for cultic purposes to celebrate the divinities, or the kings who were their stewards on earth.

From the first, distinctive characteristics, a part of the 'form' of the civilisation that was to persist so long in the Tigris–Euphrates valley, are clearly recognisable. The spirit of this art is most of all distinguished by strength and controlled energy. Human figures, muscular and sturdy, are shown in a state of arrested movement, yet they suggest an energy only awaiting release very different from the calm passivity of much Egyptian art. Even the figures on the famous 'standard' of Ur have a sturdy strength, but when it comes to the naked priests on the big vase from Uruk, the slightly later lion hunt from the same city and the militaristic 'vulture stele' of Early Dynastic Lagash, the bodily force is palpable. It appears even more forcefully in the terrifying lion-headed man of unknown provenance (Figure 1). This is the oldest known example of the composite monster that was to remain a characteristic product of the imagination of Mesopotamian artists down to the man-headed winged bulls and lions of Assyrian palace gates. Many of these weird creatures show that urge to the grotesque that has so often emerged from the depths of the human psyche.

Another recurring characteristic of the Sumerian tradition is the expression of contest and struggle. A favourite subject was the lion or other predator tearing a deer, while the hero Gilgamesh might be shown either in his battle with Enkidu or mastering wild animals; an early work shows a bull attacked by a lion and a vulture. A sense of fierce contest in, for example, Gilgamesh battling with a lion, can even be compressed into the small compass of a seal stone.

Among more purely religious subjects of divinities and worshippers there is often a comparable intensity. Even the half grotesque huge-eyed temple figures from Tell Asmar possess it and in a more naturalistic worshipper from the same site there is a profoundly troubled, brooding quality (Figure 2).

In the latest predynastic times before the unification of Egypt in about 3200 B.C. and briefly in the succeeding First Dynasty, some Egyptian relief carving shows a strong Mesopotamian influence. It is apparent in composite monsters, Gilgamesh-like masters of animals, sturdy, muscled men, on such specimens as the Gebel el-Arak knife handle and several slate palettes. It can still be seen in the weird beasts, their long necks interlinked, on the big palette of King Narmer, probably the unifier and founder of Dynasty I. Yet this same palette already shows wholly Egyptian characteristics in the static figure of the king, the division into orderly zones, the iconography – and indeed in the fine bas-relief work itself of which the Egyptians were always to be masters.

It was not long before the Mesopotamian elements had been completely absorbed and no fragment of Egyptian sculpture or painting could ever be mistaken for Sumerian or any other work. The quintessence of the art of ancient Egypt was to show all life at rest and taken out of time. In the more monumental sculptures such as seated and standing figures, positions of rest for the limbs were chosen and formalised, but even where activity is represented we see it arrested in eternity. There is a sense of inner life and power, but imbued with a calmness in utter contrast to the intense dynamism of the Sumerians. This calm power is already marvellously shown in the seated statue of Zoser, builder of the Step Pyramid, and in more advanced humanity in that of Chephren, the Dynasty IV pyramid-builder.

In my opinion too much has been made of the conceptual nature of Egyptian art, a claim based on little more than the frontal torso and other minor conventions in the reliefs. Surely, on the contrary, Egyptian artists were minutely and lovingly observant, striving to show both human beings and nature as they were. This was soon to lead to the world's first realistic portraiture. The bust of Prince Ankhaf is as full of individuality as the best Roman portrayals, so too is the plump official, Ka-aper, while the young royal couple, Mycerinus and his queen, (Figure 3) are shown as affectionate human individuals as well as divine rulers.

The artists looked at nature with the same perceptive eye for detail. Scenes in Old Kingdom tombs brings to life the plants, birds, animals, fish, and even the insects, of the Nilotic papyrus swamps. What realism as well as feeling there is in the picture of the cow lowing after her calf from the tomb of Ti at Sakkara (Figure 4).

Insufficient though they are, perhaps these pictures and references have been enough to suggest the contrast between two contemporaneous arts: the dynamism, inner drive and love of the grotesque of the Sumerians' work, the detachment and precise observation of the Egyptians'. This is made all the more significant by the manner in which the Egyptians quickly digested the strong Asian influence of their formative period so that by Early Dynastic times their art had assumed its own pure form – and from then on was usually strictly framed by tradition and its conventions.

In our mental world, where in spite of the uncertainty principle we are still so partial to cause and effect, we must look for some matter-of-fact explanations of this contrast in culture forms. To start with the most down-to-earth, we have to think of the physical nature of the lands. The wide, featureless diluvial plain of Sumeria had a fall of only one hundred feet between the present Baghdad and the Arabian Gulf. Its vast and elaborate irrigation system was forever silting up, while heavy rains in the mountains or tides and gales in the Gulf could cause floods, sudden changes in river beds and hence the ruin of cities. Although there were fertilising annual floods, their arrival was erratic and sometimes violent. The Sumerians, then, were always confronted by threats of change, disaster, a return to the chaos that was their nightmare.

How different was the Nile valley with its flood regular almost to the day, its steady fall to the Delta and the neat limits of the cultivable soil – all making irrigation easier, life more dependable and giving a sense of divinely imposed order and timelessness. The river itself, easily navigable up and down stream, was a mighty unifying force.

Turning now to the social and historical differences between the experience of the two peoples, the greatest was, of course, that while the wide Sumerian plain came to be divided between many city states frequently at war with one another, Egypt was unified as one normally peaceful kingdom from the First Cataract to the Mediterranean. Through most of the five centuries of the Old Kingdom it was a monolithic state strongly administered in the name of the all-powerful and divine Pharaoh. For this was another social and religious distinction: whereas the Sumerian cities were ruled by mere Great Men as stewards of the gods to whom their states belonged, Pharaoh was himself a god, indeed among the greatest of the pantheon.

Then there was the historical condition, largely dictated by the geographical, that while Sumeria had to share even her diluvium with the ever-hostile Elamite neighbours and was always open to raiding and invasion both from the barbarous mountain men in the north and from the Semitic nomads of Arabia, Egypt was largely isolated and secure between the western and eastern deserts, whose unorganised populations she had little difficulty in controlling.

It is easy to see how these contrasts in their lands, social structures and history could have deeply affected the psychology of Sumerians and Egyptians as peoples and as individuals. Why the Sumerians tended to be anxious, pessimistic, unstable and convinced that men had been created as slaves for somewhat unreliable deities. Even why they had no hope of permanency in this world or of continuance of happiness in another. Yet their response to being forced to live dangerously could have given them their vigour, passionate loves and hates and, in spite of some lofty moral ideals, what Professor Kramer has recognised as an 'ambitious, competitive, aggressive and seemingly far from ethical drive for . . . prestige, victory and success.' Such a psychology is surely expressed in the vital energy of their art, and with it inevitably went a deep stirring of the unconscious very evident in their mythology and in their vision of monsters and the grotesque.

The opposite effect of their relatively calm and ordered life experience upon the Egyptians scarcely needs emphasis. It cannot be doubted that it helped to form their ruling ideal of changelessness, of an eternal cycle of life under Ma'at – that power personified as a goddess and comprising order, justice and truth. What was the terrible fact that appeared to outrage changelessness under Ma'at? Why, obviously, it was death. That was why death had to be defeated through preservation of the body and all the costly and beautiful funerary practices intended to ensure that the dead still lived as they had always done, immortals returned to the endless cycle of being.

After the amazing early brilliance of the Old Kingdom, such an ideal was almost bound to lead to highly traditional arts, with a fixed symbolic meaning in every attitude, action and object. Of course there were occasional innovations and marks of genius among artists, but in general Egyptian art was far more standardised, less erratic, than the Sumerian. There was one extraordinary break with this traditionalism that I find so relevant to our present purposes that I must recall it, however hastily. I am referring to the revolution in

1 Sumerian lion-headed man, provenance unknown

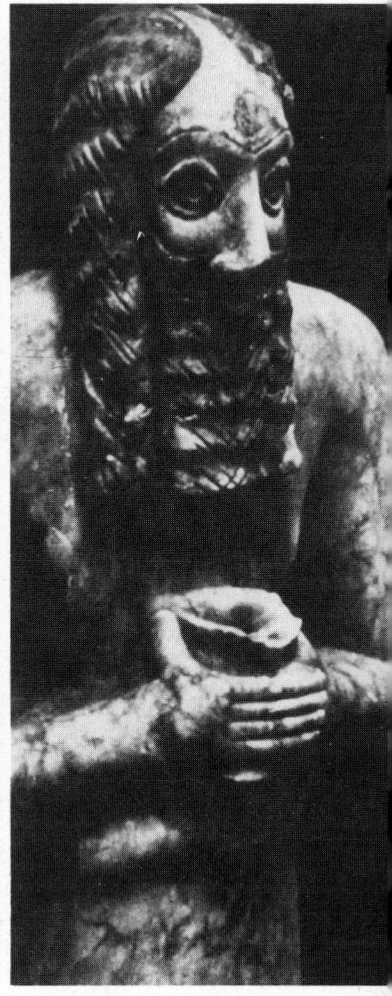

2 Sumerian worshipper, from Tell Asmar

3 Mycerinus and h
queen, from Giz
Egypt, Dynasty
2599-2571 BC

4 A cow lowing after her calf, from the tomb of Ti at Sakkara, Egypt

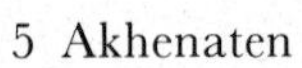

5 Akhenaten

6 Two boxers, from Akrotiri on the island of Thera, Greece

7 Swallows and lilies in a rocky landscape, from Akrotiri

8 Mayan relief head from Yaxchilan, Mexico

9 Zapotec figure of the god Xipe Totec from Monte Alban, Mexico

10 The Aztec Earth Goddess, Coatlicue, from Mexico City

almost all aspects of life brought about during the fourteenth century B.C. by the 'heretic Pharaoh', Akhenaten, (Figure 5) in partnership with his Queen, Nefertiti.

Even all the recent efforts to debunk great men and to deny their importance in history have not been able to deny the fact that it was Akhenaten himself who brought about the total revolution in religion, art and the values of life – all centred on his monotheistic worship of the Aten and his idea of 'living in truth'. An inscription records that Akhenaten's chief sculptor, Bak, had been instructed in the revolutionary art forms by the king himself. So far from an evolutionary development, the earliest phase in the visual arts was also the most extreme, a proof of its conscious imposition. Professor Aldred says, 'The mannerism of this initial phase, unlike that of any other in the history of human culture, can only owe its peculiar character to the ideas of one individual – Akhenaten . . .'

What is most significant for us in this unique cultural revolution is that although it was inspired by a single genius, it already in many ways resembled later uprisings against the restraints of a classic tradition. I am thinking particularly of our own Romantic Movement and the liberal values that have gone with it. I cannot do more than offer a shockingly over-simplified catalogue of characteristics of the Amarna revolution – named, of course, after the site of the holy city of Akhetaten that the king built away from the old cities and their gods so that his ideals could be lived to the full. Here is the catalogue.

With monotheism, the king as a god was deposed. He was the god's only prophet but as a mortal man wished to be depicted living informally with his wife and small daughters. With this must have gone much greater informality of manners at the court and in society. Nature was enjoyed for its own sake. Literature was more vernacular. There was some sense of brotherhood of peoples under the light of the Aten, accompanied if not by pacifism then at least by a lack of interest in militarism. There was a rise in the status of women and a freer, more tender, relationship between the sexes. For the first time children were freely portrayed – even in their natural naughtiness. It can be said that the whole culture was less male-dominated, more expressive of the female principle. All these qualities are reflected in the visual arts, which also change fundamentally from the old timeless, static principle to one of spontaneity and motion – an art of the moment.

Among the extreme mannerist work of the first phase of Amarna art the huge pillar statues of Akhenaten from the new Aten temple at Karnak are among the finest and most significant. They show the physical distortions that were to be emphasised in most portrayals of the royal family during the supremacy of Bak – which seems to have lasted for some eight years. There are the narrow shoulders, and wide hips, the sagging stomach and the elongated face and chin, the very full lips. The finest of these colossi also show an extraordinary intensity of facial expression, as though Pharaoh were lost in an inward mystical vision. One can well imagine the revolutionary force of such a personality given absolute kingly power.

The move to Akhetaten took place during this phase, and many of the highly animated scenes of family life in full mannerist style were found there. The royal pair are shown kissing, dandling and playing with their daughters and taking them for drives. Akhenaten himself is once shown gnawing a cutlet and Nefertiti a roasted bird. In a charming fragment of a fresco from the palace at Akhetaten two princesses are seated on cushions at their parents' feet, entirely absorbed in play with one another.

The end of the Bak art period was followed by a few years when the distortions were moderated before, in about the twelfth year of the reign, the final mature period began with the great sculptor Tuthmose. This saw the creation of the famous heads of Nefertiti, some lovely studies of the daughters as girls and young women and other masterpieces. Akhenaten was now also portrayed normally.

For the royal family to be displayed in all the intimacy of ordinary life was a complete break with tradition. This liberation lasted, particularly in tender scenes between husband and wife, into the reign of Tutankhamen. No doubt it was profoundly shocking to conservatives and was certainly ended when the soldier Horemhab took the throne. As an artistic revolution a greater significance lies in the total change from arrested movement to animation, from timelessness to an art of the fleeting moment, from unrelatedness to coherent scenes where figures are related to one another in space and also in mutual emotion.

Finally Amarna art shows the love of nature so characteristic of romantic liberalism. Realistic studies of the creatures and vegetation of the Nile were made at all times, but usually as objects of human activity. At Akhetaten nature was often painted for its own sake, free of hunters or other human intruders. In the 'paradise' water garden to the south of the city there were realistic plant paintings, while the 'green room' of the marvellous North Palace was frescoed with a continuous, free-growing papyrus thicket, sheltering lively birds of many kinds, including a diving kingfisher. Even the mangers in what was in part a royal zoological garden, were carved with appropriate animals in the animated Amarna style. Here is the expression of the simplest side of Akhenaten's ideal. As Pendlebury said, it was 'a building unique in the ancient world . . . a place where the king could watch animals and satisfy his love of nature.'

That this liberal–romantic spirit should have emerged so long ago as a coherent pattern is intensely interesting, since it suggests that the ground of the human psyche has remained constant. If the imagination rooted in the unconscious, is allowed to rise against classic formalism, its products, it seems, will have much in common.

I want now all-too-briefly to consider another culture which in many ways comes closer to that of Amarna than any other in the ancient world.

Indeed, it is possible that the Minoans of Crete had some

historical contact with Amarna. The palatial civilisation of Knossos is thought to have been destroyed either at about the time when Akhenaten's reign began (*c*. 1375 B.C.) or a generation earlier. The Cretans shared with the Atenists a delight in nature – its birds, animals, flowers – and went further than the Egyptians towards a feeling for landscape. They, too, allowed a relatively high status to women with priestesses, perhaps often princesses of the royal house, playing important roles. Probably there was a more widespread freedom between the sexes and a more open and lively sexuality. The same spirit is shown in the Cretans' indulgence in games from boxing to bull leaping, in dancing and social gaieties.

It seems that island society was altogether more lighthearted than that living in the holy city by the Nile. It had, after all, grown freely from simple agricultural origins within a lovely and secure island. There were no internal enemies waiting to destroy it.

Mrs Frankfort wrote of the Minoans as having 'the most complete acceptance of the grace of life the world has ever known,' while Arnold Hauser proclaimed 'What freedom in artistic life in contrast to the oppressive conventionalism in the rest of the Ancient-Oriental world!' This grace and freedom were rooted in a religion that had survived from the fertility cults of the old farming communities, keeping the ancient goddess on her throne, refining it for high civilisation. While undoubtedly the kings and queens of the several Cretan palaces, and ultimately the supreme rulers of Knossos, had vital religious functions, they never used theocratic power to build vast temples and their priesthoods. Minoan worship was still largely in a rustic setting with mountain-top shrines, sacred caves and trees. Even in the palaces the shrines to the goddess were simple structures for the display of the divine symbols and figurines and the reception of offerings.

Nor did the rulers seek aggrandisement by war. Towns and palaces were unfortified until the end: effort was expended on trade rather than militarism. Nor, again, did the rulers seek to immortalise themselves with grandiose tombs or statues. No piece of even life-size sculpture is known in all Crete. So in Minoan civilisation we are looking at a scene rare in history – an advanced society with wealth concentrated in royal courts, but where the presiding divinity remained female and the whole culture was dominated by the female principle.

The visual arts perfectly express this spirit. Not very much is known of them from the Old Palaces, yet already in their day the potters were producing some of the finest ceramics of all time. Although the richly coloured designs were largely non-representational – whirling spirals, stylised plant motifs and the like – already they showed the intense vitality that was to infuse the decoration and furnishing of the New Palaces and of the country mansions and town houses that were springing up over much of the island and in settlements overseas.

Most revealing and characteristic were the mural paintings that must have filled the rooms with colour and movement. Many show scenes of social and religious life: women dancing and chattering, men and women excitedly watching performances of some kind, girls and youths bull leaping together, or children at sports, such as the two little boxers at Akrotiri on Thera (Figure 6). Even more remarkable are the many frescoes devoted entirely to nature as a source of beauty and delight. Monkeys pluck flowers or leap among fantastic rocks; a cat stalks a bird (but does not catch it); dolphins, swallows, flying fish skim across the walls of private rooms. Lilies and other flowers grow formally in pots or luxuriate in a state of nature. For the first time blended colour is used to suggest atmosphere. The frescoed houses of Akrotiri, preserved below volcanic ash, have proved a revelation of the extent to which Minoan painters tackled landscape. One room was frescoed with a springtime scene of flying swallows and lilies sprouting among rocky pinnacles (Figure 7), while another had a long frieze showing a fleet of ships against a coastal background with trees, animals and a small town.

While the wall paintings perhaps best express the gaiety and colour of Minoan art, the same spirit is manifest in other forms such as carved stone vessels, small bronzes and ivory carvings, goldwork, masterpieces in faience and of course, in the superb painted pottery that included the unique marine style with octopus, nautilus, shells and coral. The staring, tentacle-brandishing octopuses seem to be intentionally humorous and the same can be said of the rollicking, drunken peasants of the Harvester Vase. Perhaps special mention should be made of the gold signet rings because they bear cult scenes with ecstatic dancing in the presence of the Goddess, offerings of lilies and opium poppies, tree worship and the like.

There is no doubt that Minoan art and all Minoan life was shot through with religious meaning, yet it appears more freely blended with the secular. Many activities, such as formal dancing, sports, the bull games, were half rite, half play and entertainment.

After this evocation of the Minoan acceptance of the grace of life within the realm of the Goddess, I want to end with an art that seems to manifest an opposite spirit. We owe to Ruth Benedict the image that each culture, each society, 'selects some segment of the arc of possible human behaviour.' It fits very well with Frankfort's conception of the 'form' of a civilisation – and, indeed, was fully accepted by him. The pre-Columbian Mesoamericans strike us as having selected the most dark and corrupt of all those possible segments. Their visual arts are once again expressive of the inner psychic life.

Even the creations of the Olmecs, who had taken the first steps towards high civilisation by the beginning of the first millennium B.C., have a sombre and more than a little sinister quality. The expression on those famous 'big heads' is morose, tormented, in spite of the baby-like features. The frequent combination of these false baby-faces with feline fangs is, to say the least, unpleasant.

It is, however, the art of the Maya, of the Toltecs and contemporary peoples of Mexico, and most obviously of the Aztecs, that can hardly fail to cause revulsion and even horror. Not only is much of the subject matter and the imaginative invention grimly full of sado-masochism, but the total style of the art seems to me to have a corrupt quality.

Classic Maya sculpture can undoubtedly be powerful and sometimes humane. For example, some of the stucco heads from Palenque and the young maize god from Copan are fine and even graceful works. On the other hand, much even of Mayan art shares in the disagreeable over-elaboration that seems to me corrupt. The relief head from a lintel at Yaxchilan (Figure 8) while wonderfully forceful has a sinister presence, perhaps enhanced by an artificial distortion of the skull. The old view that the Maya were a relatively peaceful people received a mortal blow when the series of Bonampak murals were discovered. They show prisoners cringing in fear, others decapitated, stabbed, or holding up fingers dripping with blood. The soldiers and bigwigs lined up in these scenes are in their way even more dismaying in the impression they make of a sumptuous, arrogant brutality.

There is no doubt that the Maya practised human sacrifice in many forms throughout their history. Like the Sumerians they did not think the gods were benign but had to be appeased and nourished with their own blood and palpitating hearts. They were not, however, much given to portraying these acts in their art. Nor were their divinities made to appear repulsive – though the large ceramic Death God from Tikal holding a death's head in his hands would eclipse any medieval devil in frightfulness.

At post-classic Chichen Itza in Yucatan there are horrors in plenty, the change being probably brought about by the arrival of warlike Toltecs from Mexico. Now there are bloody sacrificial scenes, renderings of the skull racks in stone probably used to support actual severed heads, while in the sacred pool (cenote) were the bones of men, women and children sacrificed by drowning.

The climax in militarism, blood sacrifice and frightfulness in art was achieved in the fifteenth century when the Aztecs conquered most of Mexico. From 10–15,000 victims, prisoners of war or purchased slaves and children, were sacrificed every year. Stinking skull racks stood near temples and there was an elaborate sacrificial equipment including richly decorated stone knives for snatching out hearts and carved bowls to receive them. Together with this slaughter went much penitential self-mutilation such as slashing the face, cutting lips and dragging cactus spines through the tongue. This masochistic aspect is well represented by the Zapotec figure of the god Xipe Totec from Monte Alban (Figure 9).

It is indeed in the iconography of their divinities that the Aztec psyche is most clearly reflected. Xipe Totec is more often shown wearing the bare, flayed skin of a prisoner – another rite of Aztec warriors. He can be taken as one aspect of Tezcatlipoca, a dark god much involved with war and sacrifice and sin (though he had a lighter side). He had lost a foot to the earth monster and often appears with the end of his tibia exposed. In a well known codex he is painted devouring the head of a sacrificial victim – one of the choice cuts that were in real death reserved for the élite. Another picture from the codex shows the fire god at the centre of the universe being fed with blood from the hand, head, ribs and leg of Tezcatlipoca; a tree fruits in the form of sacrificial knives.

The Aztecs and their contemporaries also produced many dire images of gods of the dead, particularly of Mictlantecuhtli in grim skeletal forms, expressing the notion that on the way to his sombre underworld the dead were flenched by a wind of knives. A famous statue of his consort shows her with a grinning skull head, sagging breasts and a skirt of rattlesnakes.

An eight-foot-high figure of the Earth Goddess, Coatlicue, from the Aztec capital underlying Mexico City, has been described as 'one of the most horrifying and monstrous sculptures in the world' (Figure 10). Her head is formed from the opposed heads of two rattlesnakes; her neck is a vase designed for holding sacrificial hearts, and hung round it is a necklace of hands, hearts and a skull. The shirt is woven of rattlesnakes and her feet are clawed. Could anything be further from the Minoan vision of the Goddess?

I know that value judgements of the kind I have been making are widely disapproved of, and that the religious mythologies behind the art can be called subtle and profound. A Mexican art critic has found that as a manifestation of the struggle of life and death, Coatlicue has 'a tragic and moving beauty.' I know, too, that it is easy to misjudge an entirely alien culture. For example, any extra-terrestrial visitor seeing the population of St. Sebastians in an Italian art gallery, or the bloodier types of crucifix in our Churches, might get a very false idea of Christendom. The cruelties of our Old World Bronze Age and, indeed, Roman civilisations were very great. These were sometimes recorded in a detached sort of way in works of art. There are, however, real distinctions between their cruelties and those in the New World. One, that in historical fact the Americans' utter disregard for life and suffering tended to increase, reaching a horrible climax under the Aztecs with their mass sacrifices, eating of sacrificial flesh and slaughter of tethered prisoners; their huge racks of severed heads on the public squares. The Egyptians and Sumerians did in time become sickened by the immolation of servitors, providing models in their stead, even if they continued to abuse their enemies. Second and far more significant from our point of view, the American art in itself seems to express something sinister, not only in its subject matter but in its whole style and essence. It really does look as though the imaginations that created these terrible gods and goddesses, these scenes of sacrifice, pain and humiliation, this crawling mass of unpleasant ornament, were sadistically corrupt. It appears, indeed, that the Mesoamerican peoples developed a segment of the psyche that was the most evil of those within the arc of possibility. Why? Indeed why? I will not even attempt any superficial socio-ecological explanations. One can only say that the Americans dredged up images, images tied to behavioural urges, from some corrupted depths of the unconscious.

Now that we have looked at samples of the arts of four civilisations, have we learnt anything relevant to the thinking of the Bradford Trust? I think I can only make several more or less biased statements and leave it to you to condemn or approve them. First of all there is the fact that the arts flourished long before anything properly defined as science had entered the mind of man. It is, I suppose, possible for a new, intellectual function to arise later yet side by side with the ancient image-making function and have something in common with it. Yet surely the archaic nature of the arts and the intellectual modernity of science is at least a warning against trying to run them in double harness. As I have said, the profound difference seems to be in the spontaneous, uncontrollable power of the unconscious in the genesis of works of art. Can this be 'verified' in the Bradfordian sense? One can say that its very presence *is* verification, is truth, yet it seems truth of such a very different nature from any hypothesis that can be verified by scientific experiment that I doubt if there is much meaning in using the same word for both.

Now for a second statement. Without going so far as Spengler, who spoke in simple biological terms, there is no doubt that the arts, like the rest of the 'form' of the culture or civilisation are born, mature, flourish and die in a strikingly organic manner. On the other hand technology and science, although there have been local set-backs, have overall shown continuous progress, gradual accumulation of attainment, throughout human history. This contrast must have significance for us. Bradford allows that every verification has to be taken within its own context and for us now that means within the form of each civilisation. That is well enough. Yet if the arts are always moving through something like a life cycle which has often ended in death, it does surely queer the test of truth through the public judgement. During my lifetime admiration and acceptance of the ripe classical art of the fifth century gave way to admiration of the archaic, while eyes that had seen the creations of unlettered societies only as quaint specimens were suddenly opened to their aesthetic values. The Akhenaten adventure lasted less than twenty years; what would have been condemned as sacriligious rubbish before it began was the very essence of Ma'at, or truth, and was hideous untruth once more as soon as it was over. This need not be an insuperable argument against the verification doctrine, but it is a difficulty. In particular the Bradford suggestion that if the arts sometimes appear harmful rather than life-enhancing, then these bad patches should be looked at in the perspective of evolution and appear trivial, I believe to be misleading. Because of their rise and fall, the arts cannot be looked at in that way but only on their own parabola. Arts that appear destructive may manifest the irresistible decay of a civilisation, as perhaps ours do today, or again they may belong to a 'form', such as the American, that from the first seems to rise from some mysterious corruption. Yet this art, too, would have expressed the ultimate truth to its creators.

Third statement. Our hasty study, and any study whatever, must show that throughout history the arts have flourished most, have played their most certain role in clarifying and enhancing the values of a people at a time when every aspect of their society was at its height – trade, conquering might, technological advance, religious inspiration – all under the sovereign rule of a still-trusted élite. I remember in my young and more naively liberal days feeling bitter chagrin when forced to realise that greatness in the arts nearly always coincided with power, wealth and privilege.

We have got to find our new version of these conditions – different though it must be. In the past over thousands of years much of the wealth of successful civilisations was poured into artistic-cum-religious expression while technical advance was of little importance. Now, as Bradford so rightly proclaims, the balance has tipped violently the other way and our society is weighed down with highly dangerous scientific power while remaining somewhat infantile in its imaginative life. With us the arts are no more than a decorative fringe which can be allowed a few pence from the national wealth if there is that much left over from the vast sums devoted to material hardware.

If we are now in the final stage of disintegration from which a new civilisation might spring it is evident that we should strive to redress this balance. Two immediate questions arise. How far can the arts be consciously fostered and directed towards social ends? The social realism of communist states is a warning that it cannot be done by order. On the other hand can we take the unique achievement of Akhenaten's revolution as a proof that imaginative genius *can* inspire a whole society through the age old combination of art with religious meaning? It is, I suppose we all agree, the lack of meaning in our modern lives that is reflected in our often vacant and nihilistic arts. This leads on to the second question. Bradford puts great emphasis on the role of the arts in developing a new morality. In my view although of course they can, and particularly literature and drama, be a vehicle for moral ideas, morality has nothing to do with the essence of art. So when I speak of redressing the balance, I am thinking more in terms of the balance between intellect and the imaginative powers that are fed from the unconscious. These two opposites can also be expressed in the symbols of male and female, light and earth-rich darkness – the yang and the yin. Our inheritance through Hebrews, Greeks and the Renaissance has left us, as psychic energy ebbs away, with a gross overweighting of the masculine yang. There seems to me just a faint hope that a rebirth of a new form of civilisation may come about and that in it the yin will have its due weight. It follows that the imaginative arts would then come in to their own again, and the civilisation could manifest the joyful balance and poise achieved by the Minoans, and perhaps attempted at that strange holy city of Amarna.

I know that I have tried to do too much, have been superficial and left a score of ends dangling. I only hope that I may have stirred up thoughts of agreement or of opposition and that some of the ideas I have touched on may prove to be relevant to the difficulties and possibilities confronting the work of the Richard Bradford Trust.

26th November, 1976

LECTURE IV

The Philosophy of Karl Popper

Peter Medawar

It is convenient, though (as I shall explain later) somewhat artificial, to consider the philosophy of Professor Sir Karl Popper, F.R.S., F.B.A., under two headings: on the one hand, the philosophy of history and of social science, and on the other hand the philosophy of natural science.

Popper's philosophy of history and of social science

Quite early in the nineteenth century, the recognition that science and technology were incomparably the most successful endeavours human beings had ever engaged upon, nourished the ambitious belief that the methods which had been so successful in science and technology, could be applied to human beings and to society.

It was widely hoped and believed that it would be possible to recognise and propound laws of the historic process and of social transformation – in effect, laws of historical destiny – which could have the same predictive value as the laws of physics and chemistry. When Edmond Halley predicted in 1705 that his comet would return in 1758, he was doing more than social scientists might be able to do if only they were able to formulate the laws of historical change.

For beliefs of this general kind, Popper coined the word 'historicism'. In his first publication in the English language

– a series of papers in *Economica*[1] entitled: *The Poverty of Historicism* – Popper showed that historicism was without logical foundation – was, indeed, mistaken. Popper's position in the social sciences may therefore be said to be analogous to David Hume's in the natural sciences, for Hume had caused a similar upheaval of accepted conventional thought when in much the same quiet, unassuming and lethally effective way he destroyed the foundations of empiricism by showing that the law of causality and the principle of the uniformity of nature were based not only upon logical reasoning but upon faith – upon a mere habit of expectation.

A historicist argument

I shall now give a real-life example of a fallacious argument based on historicist principles – an example which illustrates very clearly the kind of reason why Popper, in *The Poverty of Historicism*, rejected the possibility of devising predictive laws of human destiny.

What can the laws of economic determinism tell us about placing and design of factories and the dwellings that tend to grow up around them? Popper cites the following typically Marxist argument. (Unfortunately neither Sir Karl Popper, Mr Shearmur nor I, can remember the passage in Popper's writings in which this argument is discussed in detail.) The argument goes as follows:

> The source of energy is coal: therefore factories and the dwelling places that go with them must be situated near coalfields, to reduce to a minimum the economic burden of transporting coal.
>
> Coal yields power through the agency of *steam*, but as it is hardly feasible for each machine tool in a factory to be powered by its own steam engine the designer of the factory will so contrive it that a single steam engine drives a single overhead shaft or spindle, which is connected by flexible belts to the individual machine tools on the shop floor.

> Thus not only will factories be ever more concentrated in areas near coalfields, but individual factories will get larger and larger so as to get the maximum economic benefit from the operation of a single steam plant.

Unfortunately for theories of economic determinism, the predictions embodied in these arguments have not been borne out, for Marx was unable to foresee, and it would indeed have been impossible in principle for *anybody* to foresee, the advent of that electrical power which has made it possible for large numbers of small, as opposed to small numbers of large, factories to spring up all over the place, not just in the immediate neighbourhood of coalfields, because the economic argument relating to coal transport is now no longer valid. This is precisely what has happened in the light engineering works of the English Midlands, and of the State of Massachusetts.

This whole argument illustrates Popper's reasoning at the very beginning of *The Poverty of Historicism*: human history is, at least to some degree, influenced by scientific and technological ideas, and inasmuch as these ideas are intrinsically unpredictable, so one cannot predict the future course of history:

> This means we must reject the possibility of a theoretical history; that is to say, of a historical social science that would correspond to theoretical physics. There can be no scientific theory of historical development serving as a basis for historical prediction.

It is very much in keeping with Popper's philosophy of history that he is deeply opposed to revolutionary or 'holistic' remedies for whatever may be amiss with society. (Holists believe that a society is an organism with so high a degree of integration of its parts that no tinkering with them is possible.) After all, a society – if it exists at all as a society – must already have a certain organisation and working capability which makes it most unlikely that its imperfections can be remedied bringing the whole of it tumbling down through revolutionary action or – what comes to the same thing – that it can be remedied by some kind of comprehensive over-all treatment that changes society's entire structure into that of an organism of some quite different kind. The formula that has come to be especially associated with the name of Karl Popper is that of 'piecemeal social engineering'. It is not by revolutionary or holistic action that society will be improved. A multitude of things go wrong in any real society, and the important task of the social reformer is to find out in which particular way things have gone amiss and then, if possible, to put them right in such a way that he can find out whether or not his remedies have the desired effect. . . .

> The piecemeal engineer knows, like Socrates, how little he knows. He knows that we can learn only from our mistakes. Accordingly, he will make his way step by step, carefully comparing the results expected with the results achieved, and be always on the lookout for unavoidable, unwanted consequences of any reform; and he will avoid undertaking reforms of a complexity and scope which will make it impossible for him to disentangle causes and effects, and to know what he is really doing. (p. 67.)

Good real-life examples of social or economic engineering that have been embarked upon without a sufficiently anxious awareness of the possibility of untoward consequences are the cavalier attitude towards inflation of Keynsian economics, and at a less grandiose level, the Rent Acts which, though intended to produce one effect, sometimes had exactly the opposite effect.

The open society and its enemies

Anybody who dismissed the arguments of *The Poverty of Historicism* as too abstractly philosophic in character must have been strangely lacking in sensibility, because from the Spanish Civil War onwards Europe seemed to have become a test-tube for the interaction between the two great historicist doctrines which prevailed at the time – Fascism and Marxism.

According to Marxist theory, the direction of the flow of history is shaped by a struggle for supremacy between social classes – in particular between the proletariat and those who own the means of production. It was argued that this struggle would inevitably lead to a social revolution, and end in the victory of the proletariat, with the disappearance of a class stratification.

The gist of Fascism cannot be summarised in any such familiar form of words, but the high Fascism of Nazi apologists such as Alfred Rosenberg, was a form of racial or genetic élitism, which positively avowed that the advancement of mankind and the motivation of historical change were the special responsibility of a single race in the interests of which all opposition could justifiably be trampled down.

There is no need today to draw special attention to the misery and moral diminishment of Man for which the political realisations of the two great historicist doctrines have been responsibile, but in the early days of the war, and for some time after, there *was* a most urgent need for a philosopher to undertake a philosophic defence of democracy. This formidable task was discharged by Karl Popper in the work he wrote during his wartime exile in New Zealand: *The Open Society and its Enemies*. Although it is a work of great gravity and expository detail, it is also very dramatic in character: it fired and inspired some of the more thoughtful young people of the day, and, more important, equipped them to rebut the dangerous argument that an old-fashioned democracy could not combat the menace of totalitarian states without becoming in some respects totalitarian itself.

We know from Popper's brief autobiography, in which he comments on the deadly dullness of some of the lectures he endured as a secondary-school pupil in Vienna, that Popper fully appreciated the importance of an arresting or even shocking presentation.

The Open Society was dramatic and shocking because of its temperate but nevertheless lethally effective criticisms of Plato, Hegel and Marx. Popper criticises them not because they are so many sacred cows, but because 'great men make great mistakes': if civilisation is to survive we must break the habit of deference to great men. 'The aims of civilisation', he says, 'are humaneness and reasonableness, equality and freedom'. These great men, though we had been taught to admire or even revere them, had made these ambitions harder to realise.

An open society, such as our own, is a society in which disagreement and dissent, rather than being prohibited, are used as agencies of social improvement, for it is by criticising intended legislation before it becomes statutory that we may hope to discover its imperfections in time to prevent ourselves from making serious mistakes. In an open society people can flourish in all their rich and sometimes strange diversity of political opinions, ethnic origins and religious belief. In a closed society – Popper sometimes calls it a 'tribal society' – we are confined by submission to political forces or tribal observances. Only an 'open society' sets free the critical powers of man.

The Open Society was confessedly Popper's 'war-work,' and to equip himself for it and especially for his reappraisal of Plato, he set himself to study Greek again.

It was not to be expected that Marxists would greet *The Open Society* with a clamour of enthusiasm, but Popper went out of his way to ensure that a Marxist should review his book. The first review submitted to *The British Journal for the Philosophy of Science* was so laudatory that Popper, having some editorial say in the matter, proposed that the book should be sent to J.D. Bernal[2] for reviewing.

The review is most disappointing, and not all what one could have hoped from one of his remarkable intelligence. He chides Popper for misusing the word 'historicist', evidently unaware that Popper invented it, and he coins a new word of his own. Popper is a 'philagnoist' – a lover of ignorance – a description as about as plausible to call him the principal *tenore robusto* of La Scala, Milan.

Bernal makes some criticisms which make it pretty clear that he had not read *The Poverty of Historicism*. In one of the criticisms, which shows that he had not read *The Open Society* as attentively as a reviewer should, he says that Popper accuses Marx of Utopianism, whereas in reality Popper, intent upon more serious criticisms, describes this misconception of Marxian theory as 'vulgar Marxism', part of a popular misapprehension of Marxist teachings.

From time to time one still meets people who have not read *The Open Society and Its Enemies* and I now accordingly beseech them to do so. Its arguments are still relevant to the modern world, and will continue to be so for as long as democracy is under threat from any form of tribalism.

The philosophy of science

Just as Popper's *Philosophy of History and of the Social Sciences* grew out of his dissatisfaction with the complacently mistaken doctrines of historicism, so his *Philosophy of Science* could be said to have grown out of his criticism of the prevailing and orthodox opinion about how scientists go about their work to enlarge human knowledge: the opinion that scientists work by a special process known as *induction*.

Inductivism is not one but a complex of different notions, which hang together and are founded on the principle that the scientist is essentially a man who observes nature clearly and intently and without any preconception in mind. The record of an observation is an 'observant-statement.' Science therefore consists of a majestic pile of observation-statements, and all scientific laws are compounded of these observation-statements by a system of rules of discovery – as John Stuart Mill called them – or more generally (and to make it sound more sophisticated) by what we may call 'a calculus of discovery'. The high priest of induction, John

Stuart Mill, explained it in terms which, though not intended to do so, undermine it as effectively as David Hume had undermined it in his *Treatise of Human Nature* a hundred years before.

Mill said of induction ('that great mental operation') that it was a process of inference which

> proceeds from the known to the unknown . . . any process in which what seems the conclusion is no wider than the premises from which it is drawn does not fall within the meaning of the term.

As Mill also believed that induction was a logically rigorous process, so he is, in effect, telling us that by some rigorous act of mind it is possible to increase the empirical content of the assumptions it starts from. I don't think Mill quite understood what an audacious, even outrageous statement he had made, for if we could enlarge empirical information by an act of mind we scientists could conduct our research in a supine position with the eyes half closed.

Much else is wrong with induction. Using the word 'induction' now to refer to any scheme of thought which purports to show that general statements can be compounded of particular observation-statements, induction has no explanation to offer of two very familiar elements in scientific thought and discovery: error and luck, for why should we ever fall into error[3] if we observe things correctly and operate a calculus such as Mill's according to the rules? As to luck, which all scientists benefit from, how can we *know* when we are being lucky except in terms of the fulfilment of some specific prior expectation?

Induction, moreover, is prey to a number of paradoxes, the existence of which is a sure sign that something, somewhere, is logically amiss. The most amusing we may call 'the paradox of the old black boot': philosophers are not very imaginative when asked for examples of an inductive generalisation, and they are very apt to say: 'All swans are white', so let us proceed from there. If all swans are white, then all non-white objects are non-swans. (That is a cast iron logical inference.) This logical prediction is borne out by the discovery of an old black boot: it is not a swan, and it is not white, so our confidence in the generalisation that 'all swans are white' is appreciably strengthened.

Popper's solution of the problem of induction

Immanuel Kant, the foremost philosopher of his day, thought to remedy the shortcomings of empiricism by the attempted discovery and formulation of *a priori* knowledge – that is to say, of knowledge independent of all experience, and this great ambition is the programme of his *Critique of Pure Reason*. Popper's solution is to abandon altogether the search for apodeictic certainty – for truth that is demonstrably certain and therefore beyond criticism – and he proposes instead that all scientific knowledge is conjectural in character.

In Popper's view the generative act in scientific discovery or in the solution of a problem is the formulation of an hypothesis, i.e., an imaginative conjecture about what the truth of the matter might be. An hypothesis is a sort of draft law or guess about what the world – or some particularly interesting part of it – may be like, or in a wider sense it may be a mechanical invention which we can think of as a solid or embodied hypothesis.

In the outcome science is not a collection of facts or of unquestionable generalisations, but a logically connected network of hypotheses which represent our current opinion about what the real world is like.

Most of the day-to-day business of science consists not of hunting for facts as an inductivist might suppose, but of testing hypotheses, that is, seeing if they stand up to the test of real life or – if inventions – to see whether or not they work. Acts undertaken to test a hypothesis are referred to as 'experiments'.

What is being tested in an experiment is the logical implications of the hypothesis, i.e. the logical consequences of accepting a hypothesis. A well designed and technically successful experiment will yield results of two different kinds: the experimental results may square with the hypothesis, or they may be inconsistent with it.

If the results square with the implications of the hypothesis, then the scientist takes heart and begins to hope that he is thinking on the right lines; he will then, if he has any sense, expose his hypothesis to still more exacting experimental tests. The riskier the hypothesis, i.e. the more 'way-out' or unlikely it seems to be in terms of current expectations, the more reassured the experimenter will be if it stands up to experiment; but no matter how often the hypothesis is confirmed – no matter how many apples fall downwards instead of upwards – the hypothesis embodying the Newtonian gravitational scheme cannot be said to have been *proved to be true*. Any hypothesis is still *sub judice* and may conceivably be supplanted by a different hypothesis later on. Many people find this element of Popper's philosophy rather disagreeable, and are inclined to think that some scientific generalisations are true beyond any serious possibility of further question.

Popper does not agree: in the whole history of science no theory has seemed more secure, more naturally and essentially right and less likely to be controverted than the body of theory that makes up Newton's celestial mechanics, but even this has been questioned and supplanted by the more general theory of Einstein. One of our foremost theoretical physicists, Professor Sir Herman Bondi, has said of Newton's theory: 'We may certainly speak of disproof now.'

The second possibility we may envisage is that the results of a carefully designed and well-executed experiment controvert the implications of the hypothesis. In real life this does not mean that the hypothesis is promptly abandoned; but some reconsideration is certainly called for which may, in the extreme case, amount to an abandonment of an hypothesis but is more likely to involve some modification of its current form. If, however, certain theories or observations are assumed to be true, then we may sometimes speak of the outright falsification of an hypothesis. The qualification: if certain theories and observations are assumed to be true is important, for falsification is not itself immune to error and something must always be assumed to be true if anything is to be shown to be false.

The demarcation problem

In one way falsification plays a specially important part in Popper's scientific philosophy. Suppose we put to ourselves the general question of what distinguishes statements which belong to the world of science and of commonsense (we need make no distinction between the two) from metaphysical or fanciful statements?

Popper's answer would be that statements belonging to the world of discourse of science and commonsense are in principle falsifiable and it must be possible in principle to envisage what steps we could take to test the statement and so maybe to find it wanting.

Popper does *not* go on to say (as in incognate circumstances logical positivists would have said) that the criterion of falsifiability-in-principle distinguishes scientific or commonsensical statements from metaphysics or nonsense. On the contrary: the line of demarcation is between the statements belonging to the world of science and commonsense and statements belonging to some other world of discourse; so far from being nonsensical, metaphysical statements may lie on the pathway towards truth and may sometimes be conducive to the discovery of the truth.

To my mind the great strength of Karl Popper's conception of the scientific process is that it is realistic – it gives a pretty fair picture of what actually goes on in real-life laboratories.

In real laboratories there is no constant clamour of affirmation or denial. We are all very conscious of being engaged in an exploratory process as we cautiously grope our way forwards by the method which has come to be summed up by the now familiar cliché of *conjecture and refutation*.

Some people think the idea that science is conjectural in character in some way diminishes science and those who practise it; but to my mind nothing could be more diminishing than the idea that the scientist is a collector and classifier of facts, a man who cranks some well-oiled machine of discovery. Popper's conception of science is, in my opinion, a liberating one; I feel enlarged, not diminished, by the thought that any truth begins life as an imaginative preconception of what the truth might be, for it puts me on the same footing as all other people who use the imaginative faculty. I feel that what distinguishes the natural scientist from laymen is that we scientists have the most elaborate critical apparatus for testing ideas: we need not persist in error if we are determined not to do so.

The communications engineer would have no difficulty in recognising Popper's formulation, the elements of a control process or a steering (cybernetic) process which accords exactly with the idea that the scientist is finding his way about the world. 'Feedback', the fundamental stratagem in all control systems, is the control of a performance by the consequences of the act performed. If we regard the logical consequences we test by experiment as the logical 'output' of an hypothesis, then it is clear that the experiment which may cause an hypothesis to be modified or even in an extreme case abandoned, gives a textbook example of the phenomenon of negative feedback.

No scientist thinks of himself as a man of facts and calculation, Popper puts it thus:

> It is not his *possession* of knowledge that makes the man of science, but his persistent and relentlessly critical search for the truth.

Peter Medawar

I said right at the outset that though we can consider Popper's philosophy under two principal headings – the philosophy of history and of the social sciences, and the philosophy of the natural sciences – nevertheless I also said that the distinction was rather an artificial one. Why? – What is it that they have in common?

What they have in common is the element that pervades the whole of Popper's philosophy: the recognition that human designs and human schemes of thought are very often (perhaps more often than not) mistaken, and that the safest way to proceed is to identify and learn from our mistakes and learn always to do better next time. In this way, Popper believes, as I also believe, the world can be made a better place to live in.

4th March, 1977

The works of Karl Popper

There is a complete bibliography of the works of Karl Popper in the second part of *The Philosophy of Karl Popper*, 1974, edited by Arthur Schilpp, Open Court Publishing Company.

The Poverty of Historicism (1957, Third Edition, revised, Routledge, London) may be regarded as a prolegomenon to *The Open Society and Its Enemies* (1957, revised and enlarged edition, Routledge, London).

Popper's *Philosophy of the Natural Sciences* was first expounded in *Logik der Forschung* in 1933; translated into English as *The Logic of Scientific Discovery* (1972, third English edition, revised and enlarged, Hutchinson). Many of the matters discussed in *The Logic of Scientific Discovery* are taken up again and expanded and clarified in *Conjectures and Refutations: The Growth of Scientific Knowledge* (1972, Fourth Edition, revised and enlarged Routledge and Kegan Paul, London). Lastly, in 1982 was published *Quantum Theory and the Schism in Physics*, one of the three volumes of the Postscript to *The Logic of Scientific Discovery*.

Because it deals with matters not touched upon in the text – the objective existence of the world of thought – particular attention should be drawn to Popper's *Objective Thought, an Evolutionary Approach* (1972, Oxford University Press).

References

1. K.R. Popper, *Economica* vol. XI Nos. 5, 42, and 43 (1944); and vol. XII, No. 46 (1945), Third edition, Routledge and Kegan Paul Ltd; London. (1957).
2. J.D. Bernal, Has History a Meaning? *Brit. J. Philos. Sci.*, **6**, pp. 104–167. (1956).
3. K.R. Popper, On the Sources of Knowledge and of Ignorance. In *Conjectures and Refutations*. Routledge and Kegan Paul Ltd; London. (1963).

LECTURE V

Some Facts and Theories Regarding Research on the Brain

David Samuel

'Science', it has been said,[1] 'has two major objectives – to discover and describe phenomena in the world of our experience, and to establish general principles by means of which they can be explained and perhaps predicted'. A great deal has been written on the way in which these objectives are attained. Various schools of the Philosophy of Science have analysed and debated the Scientific Method in an attempt to understand how scientists reason or are presumed to reason. Much of the debate surrounds the formulation and function of scientific concepts and the definitions and terminology that are used. Some scholars have adopted an historical approach; others have based their analysis on logic or on semantics.

However, few practising scientists are aware of these debates. They operate within a complex network of hypotheses, observations and interpretations of their own and are often even prone to personal prejudices and to wishful thinking. The language of 'working' scientists is often imprecise and their methods of communication – the scientific papers in which their daily labours, hopes, dilemmas and successes are reported – never reveal the actual order in which they tackle problems or disclose the real mental processes undergone in their research.

I shall, therefore, dwell only briefly on the Scientific Method *in the abstract* and then proceed to illustrate the process of research using two examples from studies in which I have personally been involved. This will, I hope, demonstrate some of the successes and some of the failures of the Scientific approach in achieving its objectives.

The pursuit of Science, and indeed of most of the Arts, arises from a basic human motivation to make order out of chaos. This fundamental urge of mankind has existed since recorded history. It is interesting that in the quest for order and meaning, and the establishment of causal relationships, a hierarchy can be seen in the sequence in which scientific disciplines were developed. Deliberate observations and the search for order started with astronomy, were followed by physics, then by chemistry and, as nearly everyone will agree, this century has become the golden era of biology. In the 1970s, as I shall discuss later, we are also witnessing the rapid expansion of an entirely new scientific discipline – which, for want of a better name, is called the Neurosciences, which includes all research on the nervous system, the brain and on behaviour. It is relevant, also, that this hierarchy of disciplines started in those areas that are most distant from human involvement, with astronomy and with physics, areas of activity in which the observer plays no role in what is observed. At the other end of the spectrum are biology, anthropology, ethology and psychology, the sciences of human and animal function in which investigators often use themselves as subjects and are therefore both observers and observed. Perhaps this is because that which is more remote from us is easier for us to examine objectively.

Personal involvement in the subject under investigation has, on the other hand, introduced many methodological difficulties to psychology and behavioural research and has made it harder to separate fact from opinion.

Basic scientific methodology starts with the collation of facts – and with an attempt to invest these with meaning.

This data or evidence can be either collated by observers, or is the result of deliberately planned experiments; both sources of information obeying a basic rule, that of reproducibility. This means that the data or evidence must be *public*, that is, it can be secured by any competent observers or experimenters, but does not depend on who they are or on what they might anticipate by way of results. Reproducibility *does* depend, however, and to a high degree, on the clarity, uniformity and completeness of the descriptions of data and how they were obtained. The second step is correlation of these facts into a working hypothesis. Much has been said and written about which comes first – collation or hypothesis. Karl Popper[2] and others have stated that scientists do not start from random or chance observations but always with the investigation of problems based on some assumptions or *a priori* theory. This may be so in well-established fields, such as modern physics and chemistry, but in the newer disciplines of today chance observations and experiments made in another connexion frequently provide the initial data. Historically, even astronomy was first studied to aid navigation on land and sea, or for predicting the correct times for sowing – and it was only when enough facts had accumulated after centuries of observation, that bold men were tempted to formulate hypotheses in order to explain the laws of motion of the planets and the stars, and eventually to propose the heliocentricity of the solar system.

In newer fields of endeavour, such as biochemistry, which was created by dividing chemistry and combining it with part of biology, the data carried over from the parent disciplines were obtained with other purposes in mind. The study of the causes of things had to be preceded by the study of things caused. On the basis of patterns or regularities in these 'things', hypothesis is generated – which must account for all, or as much as possible, of the data. The process of framing a theoretical structure based on these relationships, is then almost simultaneously accompanied by an interpretation – in the form of a testable theory.

I will not discuss the arguments concerning the relative importance of the verification or the falsification of a hypothesis. It has even been said that a hypothesis that cannot be falsified belongs to a realm of discourse that is not science. Be that as it may, it is certain that few hypotheses in the classical sciences can be tested (or refuted) in isolation. They are almost always related to a previously existing hypothesis – such as the basis on which the measuring instruments operate. In all disciplines, the important feature is that conformity to patterns or regularities make it possible to test a hypothesis by prediction – that is for anticipated new relationships or facts. The crucial step is when the production of new facts causes a hypothesis to be modified – or on occasion abandoned. Here the burden of proof rests on those who plan and perform the experiments or observe the phenomena. They are not always unprejudiced, and in the urge to interpret or impose order, there is always the possibility of a systematic error, through the systematic misinterpretation of facts.

A case in point was the intense and renewed interest about ten years ago in a basic area of chemistry – the structure of water. This was due to the claims of Russian scientists to have produced a new form of polymeric water, popularly called 'polywater', by distilling ordinary water very slowly into fine glass capillaries. Many theories and hypothetical structures, such as long chains or networks of molecules, were proposed to account for its high boiling-point and other strange properties. However after years of debate, more careful chemical analysis showed that this 'anomalous' water was only ordinary water, contaminated with substances washed out from the silica tubes – just as Robert Robinson[3] had suggested very early in the controversy. New facts, in the form of more rigorous chemical analyses, had caused all the ingenious hypotheses based on the existence of 'polywater' to be instantly discarded. This is an unusual case in chemistry and demonstrates, as Francis Bacon[4] once said, that 'Truth emerges more readily from error than from confusion'.

More often, of course, scientific research progresses in a more orderly manner; experiments confirm the basic premise of each hypothesis, although new facts, which do not quite fit, can cause it to be modified. Let me give another example also related to the structure of water. Water, H_2O, as we all know, is composed of hydrogen and oxygen. Ever since its chemical structure was established, it was assumed that water was a pure substance with the same properties – whatever its source, and that all the atoms of oxygen and of hydrogen in it have the same size and weight. However, in 1929, in the course of observing the line spectrum of sunlight, faint shadows were seen near the lines attributed to the oxygen in the atmosphere. It was suggested that these might be due to two new forms or 'isotopes' of oxygen – atoms with virtually the same chemical properties but weighing very slightly more. This observation, and the subsequent hypothesis that these isotopes might exist in *other* materials containing oxygen, including water, led to the prediction that they could be extracted or separated by some physical process.

In the course of trying to do so, by the careful distillation of water, Harold Urey[5] discovered that hydrogen also had an isotope – heavy hydrogen (or deuterium) – nowadays far better known than heavy oxygen. Urey showed that water is indeed a mixture of molecules, with different amounts of isotopes of oxygen and hydrogen in it, depending on its source or previous history. This discovery, it turned out, has been of very great value in many areas of research from geology to biology and to the operation of nuclear reactors. It also led to two important revisions in fundamental scientific tenets. First, that due to variations in its isotopic content, the density of water was not constant, and it could no longer be used as the basis for determination of the fundamental unit of physics, the gram; secondly that, since the proportions of the three isotopes of oxygen, in all its compounds, varies, this element could no longer be used as the basic unit for determining a fundamental property of atoms – their atomic weights. Oxygen was, in fact, eventually replaced by carbon as the standard of atomic weight.

This is just one illustration, in an area in which I myself have some interest, showing that the presence of isotopes of light elements found in the telescope of an American observatory, resulted in major revisions in physics and chemistry. It is an excellent example of the successful application of the scientific method – the correlation of facts from different sources to generate a new hypothesis. It is also a measure of the imagination and creativity of a scientist, who – in seeing unusual relationships or by matching diverse facts – is led to propose a new hypothesis. This ability to see connexions beyond those of simple logic, is where perhaps the scientific method most often approaches that of an art. Insight and inspiration often play as important a role as does logical step-wise reasoning. Einstein[6] once wrote that 'there is no logical way to the discovery of these elemental laws [of physics]. There is only the way of intuition, which is helped by a feeling for the order lying behind the appearance.' Nevertheless, it must be remembered that although intuition can hasten the process, and even bring about certain discoveries much earlier than otherwise, it is *not*, in a strict sense, indispensable.

First and foremost, it is reason that plays a dominant role in extending scientific knowledge, as you will see in the two examples of brain research that I shall discuss. In each case I shall begin by summarising the facts that led in the mid-1960s to a relevant working hypothesis. I have chosen this arbitrary starting point some ten years ago – simply because my interests turned towards these problems at that time. I shall then describe what attempts were made by various groups of scientists, over the past decade, to find further evidence, that would either support or disprove these early hypotheses, outline what modifications were introduced, and then summarise where we stand today (1977).

The first of these areas of research is the problem of Memory. What *is* memory? *How* does it happen? *Where* is it? These have been central questions in philosophy, in psychology and in biology, for a very long time. Even Plato proffered a suggestion that 'there exists in the mind of man a block of wax, and that we remember what is imprinted on it as long as the image lasts, but when the image is effaced, we forget'.

Since Plato's time, a vast behavioural literature has been written both on learning and on forgetting. The study of the acquisition and of the loss of responses has indeed been one of the basic methods by means of which memory can be studied. The data accumulated essentially from two extensive lines of investigation; studies on human memory, that ranged from the probing of language and linguistics in adults, and the intellectual development of children, all the way to the abnormal aspects of memory due, for the most part to trauma, such as amnesia and aphasia. The basic technique used to ascertain what is remembered and what forgotten by humans, is one of question-and-answer. Questions (and answers) to the child, to the amnesic patient, to the cave-dwelling tribe or to one-self – as Hermann Ebbinghaus[7] did nearly a hundred years ago in the course of measuring his own ability to recall lists of nonsense syllables – thereby establishing many of the basic ideas on memorising.

The other main source of facts on learning and memory came from animal experiments; half a century of research on the improvement in the performance of rats in a maze, and on the effect of surgical lesions or drugs on their response to stimuli. It also included a great deal of work on the behaviour of many other animals from the octopus and the pigeon to the great apes. The basic difference between the two approaches is due to the fact that animals do not talk and consequently this type of behavioural research is based largely on the observation of their movements, and responses to stimuli.

The data on defects in human memory were also rerived, to a great extent, from tests on patients following injury or surgery. Both lines of investigation, in animals and in people, gradually approached one another, and were eventually integrated in the 1950s. Donald Hebb[8] in the introduction to his well-known book *The Organisation of Behaviour*, pointed out that the task of the psychologist, the task of understanding behaviour and reducing the vagaries of human thought to a mechanical process of cause and effect, is a more difficult one than that of any other scientist. 'There is a long way to go,' he wrote, 'before we can speak of understanding the principles of behaviour to the degree that we understand the principles of chemical reaction'. He also made an attempt to describe in physiological terms how information is stored in the brain and by what mechanism memory is retrieved or lost.

By the 1960s, when I began to be interested in the problem of memory and learning, there was a sense of optimism in the air. This was largely due to the great success of applying chemistry to cell biology. It was hoped that a similar link might be forged between molecular biology and the rapidly growing science of behaviour. Many psychologists, but it must be admitted fewer biochemists, considered that learning, which can be defined as 'changes in behaviour due to experience', could be related as Hebb had vaguely suggested, to some parallel change in the molecular or cellular structure of the brain. Once this relationship was established it was reasoned by the optimists that the basic questions on memory, on learning and forgetting would at last be answered.

Let us look at just what was known by the mid 1960s. Years of painstaking research by Karl Lashley,[9] and others on the effect of brain lesions on the behaviour of rats in a maze had failed to locate the so-called 'engram' – the almost mythical anatomical localisation of memory. In fact Lashley eventually came to the half-serious conclusion that 'after looking at the brain for so long, learning was just not possible'. Nevertheless a great deal *had* been learned in the course of two or three decades on the basic rules by which both men and animals remember. Successive generations of psychologists had succeeded in unravelling some of the complex effects of repetition and of interference on learning, of motivation and of reward. At the same time neurophysiologists were establishing the general principles by means of which nerve cells communicate with one another: by bursts of electrical activity traversing rapidly along connecting fibres. Combining physiology and psychology led to the suggestion that information was constantly pouring into the brain and was processed on *two* time scales. First, as short-term memory, which lasts from seconds to a few minutes, still little understood but probably consisting of patterns of electrochemical activity during which the preliminary processing takes place. In humans, at least, this may be the time in which the initial coding occurs, when incoming information is converted into silent words and classified by association. This type of memory is easily lost – destroyed in the course of minutes by interference by subsequent incoming information or by shock or physical trauma to the brain. Then by a process not clearly understood, which is called 'consolidation', the transient trace is converted into a more permanent record, long-term memory. From this relatively permanent store, what has been learned can be located and retrieved after the passage of many days, years or even a lifetime. However, even in the hands of the most talented anatomists or physiologists of that time, no measurable difference could be found in the central nervous system of an animal that learned, i.e. that has been trained to perform some task. Obviously no change in the overall size or shape of the brain or the number of cells can occur fast enough to account for consolidation nor provide any reasonable explanation for the permanent record. The search was, therefore, extended to lower levels of organisation, to the connexions between nerve cells and to the structure of molecules inside them.

What was known about the cellular structure and chemical composition of the brain in the 1960s? It was known that the brain is a most intricate organ containing many diverse regions – each with a complex specialised function.

Although many of the functions of the central nervous system were fairly well known, the contributions of biochemistry, of neurochemistry, were still modest at that time. The great success of molecular biology had led to a remarkable understanding of the mechanism of growth and reproduction in simple cells. This was then naturally applied to the brain, which was also found to contain tens of thousands of different molecules, both big and small. The small molecules were easily identified; water, of course, various salts and many of the simple well-known organic compounds required for the basic energy and maintenance of a living system. What was found to be unique to the nervous system were a dozen neuro-transmitters – special 'messenger' molecules, which when crossing minute synaptic gaps at the appropriate time, transmit nerve impulses from one cell to another.

There are billions of nerve cells in a single brain and synaptic transmission is considered to be the basic method of controlling the flow of impulses throughout the nervous system. This is the giant network that has to cope, sort and eventually record all incoming information. How much is there to record? The mathematician, J. Von Neumann[10] once estimated that an average human being may have to store at least a billion billion (10^{18}) items of information in the course of a lifetime! He may have overestimated it by several orders of magnitude, but no one could see any way of relating this number to any change in the number of cells or the structure of neuro-transmitters used.

Let us look more closely at this flood of information. Where does it come from? First of all, there is the information that is genetic – the pre-natally determined memory of a whole species and its past, coded in the double helix of deoxyribonucleic acid (DNA). These long, chain-like molecules are the messengers delivering information from one generation to another, periodically switching on or off the synthesis of thousands of different proteins and other materials, to build barriers and membranes, or to manufacture everything else that the brain needs or uses. It is on the basis of the information in DNA that the brain, and indeed the whole organism, is constructed. It is, in fact, as Colin Blackmore has recently said, the 'recipe' for an organism. But no two cakes are the same just as no two brains are the same. In the strands of this DNA, it is also believed, are deposited the codes that establish some of the more basic aspects of behaviour, those of the autonomic system and the involuntary reflexes, over which we have little or no control.

The second source of information enters the brain from the outside world, from the moment of birth (or perhaps even sooner) through the constant interaction of the organism with its environment. A stream of information, processed by the senses, travels to the brain to cause gradual non-specific changes in patterns of behaviour, forever changing and refining inherited characteristics. It is believed that specific behaviour is also altered in this manner, by input in the form of learning: these are the changes that occur as the result of experience, the memory for which some chemical correlate was being sought.

Of all the possibilities of material changes in the brain that were considered a decade ago for the permanent coding of acquired information, two seemed most likely: changes in the sequence of units in some very long-chain molecules (perhaps proteins, RNA or other macromolecules) or changes in the synapses between nerve cells, thereby perhaps altering the connexions between different areas of the brain. In the ensuing 10 years, a very large number of experiments were performed, in order to find evidence for or against each of these hypotheses. Owing to the structural and chemical complexity of the brain, no single approach could be used, so that diversity of methods led, on occasion, to inconclusive or even conflicting findings – and I shall only summarise the more consistent results.

It was found that after a variety of learning experiments in many animals, from salamanders to monkeys, few, if any new molecules were formed capable of carrying a code. The few such molecules that were isolated seemed insufficient for retaining any specific information. This seemed to dispose of the first hypothesis – of a molecular code for memory with a substance for each thing learned. Nevertheless increased amounts or turnover of such molecules already present were consistently reported. And, indeed, when the renewed synthesis of these molecules was prevented by injecting pharmacological drugs into the brain, animals were not capable of learning. These experiments seemed to indicate, that some change in the turnover of molecules was required for strengthening the contacts between nerve cells in the synapses. It is suggested that animals and humans are born with a network of many alternate neural pathways in their brains established by coded instructions from DNA. As soon as the brain starts to receive impressions from the outside world, in the form of interaction with the environment or learning, some pathways are reinforced and others wither. The new pathways, established by repetition or constant use, become more and more permanent, representing in a way the engram from which the original information can be reconstructed or retrieved.

The idea of specific synaptic changes as a basis for memory is not really new and was in fact first suggested by Eugenio Tanzi and I. Fattie in 1893[11] but only recently has there been some experimental evidence for it. The size of some synapses, as determined by the electron microscope, in experienced animals is statistically larger than in naive untaught controls. Other experiments have shown that synapses that are blocked by specific antibodies to their membranes prevent recall of what has already been learned. All these findings give some weight to the synaptic facilitation hypothesis, and provide a mechanism for recording information, but they do not explicitly explain the logic of the system.

The questions remain: how does so much information get sorted and stored so efficiently and for so long? How is one thing remembered while another is forgotten? A number of models of the brain, based on analogy, have been suggested – each pointing to a different device to store information. Various computer analogs have been suggested which can account for both the volume and the speed with which information is processed. Holography, the reconstitution of three-dimensional pictures by laser beams, can mimic the absence of localisation of memory – the absence of an engram. Electrical circuit models have also been suggested, some even with the capability of rudimentary learning. But these models have little to contribute except analogy, and analogy proves nothing. Even when refinements such as 'filters' (to account for variations of unknown origin in input or output), multiple 'stores' (to which there is greater or lesser access) and 'channels' (to localise the flow of information) are added, we are still left without a satisfactory theory.

The synaptic facilitation hypothesis of the 1960s still stands by default in 1977, but so far, without much experimental support. Its real virtue is that it has, like all good hypotheses, drawn attention to the directions in which further experimental work might be done. For instance, simpler systems for both behavioural and biochemical study are being sought; organisms however lowly, capable of quantifiable, reproducible learning, with as few brain cells as possible. It has, however, unfortunately been found that the simpler the system, the more diffuse and undefinable are the changes in its behaviour, so the search for the right animal goes on. The second line of attack is to refine the techniques needed for determining changes in a learning brain. These changes are apparently too fast or too small to be detected by most of the methods available at the present time. So memory research is again at the stage of gathering data, waiting for the moment to modify or extend the existing hypothesis. Until then, one can only guess that the recording mechanism may not be unitary but rely on statistics, since it must involve many alternate pathways and ensembles of cells and that the answer will certainly not be as simple as a double helix.

In summary, the biological hypothesis of memory, while compatible with certain biochemical data, has not yet established any new principles. Since progress has been so slow and the going so hard, many of those concerned with brain research in the past decade have diverted their attention to other problems in the neurosciences. Some were content to concentrate on techniques or simpler problems, whilst others were attracted to broader questions which, nevertheless, might be easier to answer.

One such problem is how a brain is formed each time with such apparent accuracy and reproducibility. It has been said in this connexion that the human brain is the most highly organised three pounds of matter in the universe. The question is: how do so many millions of nerve cells find each other and link up in what is, fortunately, almost always the correct pattern? This is of great interest to all those interested either in evolution or in development. Rather than discuss this research in detail I will merely point out that again the trend has been to turn to simple systems – an example of a well-tried method by which scientists overcome difficulties.

The mammalian brain is obviously a highly complex network with an overwhelming number of cells, so those interested in 'nerve cell recognition' use models, such as the brain of a chick embryo or a few hundred nerve cells in a tissue culture dish. Such simple experiments may help distinguish between two possibilities: that cells recognise one another by chemical markers on their surfaces; or that their connecting fibres follow a gradient of some kind to the right place. No one knows yet which of the two is correct.

Many neuroscientists were, and still are, attracted to problems relating to another most intriguing aspect of brain research: the breakdown of normal mental functions. Millions of people throughout the world suffer from schizophrenia, manic-depression and many other related neurological and emotional diseases. These are not rare afflictions and in many countries account for half of all the hospitalised patients. Owing to the reluctance of the public to face these problems until very recently an aura of secrecy, embarrassment and fear has surrounded victims of these disorders, frequently preventing open discussion and hampering scientific investigation.

However, during the past two decades the increasing use of new psycho-active drugs has reduced the twin feelings of helplessness and hopelessness of patients, families and physicians. Still, the chances are still estimated to be between two to three in a hundred that any one person will suffer from severe schizophrenia or serious depression in his or her lifetime. Mental illnesses obviously constitute a major social problem, the loss of valuable human potential and immense public expenditure. Today, with the growing concern for human welfare, psychiatrists and neurologists, together with pharmacologists and neurochemists, have joined hands in a world-wide attempt to understand the causes and help in the treatment of mental disorders. But there are many pitfalls in trying to use the straightforward scientific method. First the spectrum of the symptoms of each disorder is very broad, making clear-cut diagnosis and classification difficult. There has in fact been a great deal of discussion as to whether it is the individual, or society at large, that is out-of-step. There is also the often vociferously stated view that mental illness is a myth, that psychiatry is a pseudo-science, and that attempts to describe certain behaviour as abnormal and prescribe treatment to correct it – without coming to grips with social and ethical problems first are doomed to failure. And it is true that the problem of the meaning of 'normalcy' has not been resolved. Is it a statistical term, and if so, where does the borderline lie? What is the difference between eccentricity and insanity? In ancient times people who were 'different' were usually subjected to ridicule, persecution and even death. Up to the end of the Second World War the usual treatment for schizophrenia and manic-depression was physical restraint – straightjackets and padded cells. Sometimes shock of various kinds was tried and even irreversible surgery of the frontal lobes. Some, including Sigmund Freud, had suggested even at the turn of the century that a physiological cause for 'psychoses' might one day be found, but 30 years of post-mortem investigations had failed to reveal any consistent or reliably measurable changes in the brain of mentally ill patients. This was of course partly due to the crudeness of the tools available – the light microscope and gross biochemical analysis of brain tissue. However, by the 1930s a few far-sighted biochemists, such as J.H. Quastel[12] then working at a mental hospital in Cardiff, suggested that a connexion between madness and biochemical processes in the brain was likely. But the practical difficulties in obtaining reliable chemical data on the nervous system were even more difficult than for research on memory.

It has taken a long time for sufficient indirect biochemical data to accumulate – the evidence required for a hypothesis. More attention began to be paid to the role of some of the smaller molecules in the brain – the neuro-transmitters, which had been largely ignored in research on memory. In particular, the biogenic amines – of which dopamine is perhaps best known – were studied extensively, particularly their mode of formation and breakdown. As soon as the amines are used their breakdown products or metabolites escape from nerve terminals, pass into the blood stream, or flow down the spine and are ultimately excreted in urine. By the late 1960s the methods of chemical analysis were so improved that it was found that the minute quantities of such metabolites excreted by disturbed patients differed, at least statistically, from the levels in controls – people of the same age, with similar levels of activity etc. By then, new and revolutionary forms of therapy for mental illness had become available. A decade earlier, a chance observation of the tranquilising effects of phenothiazine derivatives had led to their widespread use in treating various forms of schizophrenia and calming severe agitation. Other drugs were soon found which alleviated anxiety, mania and depression. The use of therapeutically effective chemicals drew attention to a possible link between disorders of mood and the chemistry of the brain. Some of these drugs, although relieving the symptoms of mental disorders, caused tremors and other hallmarks of a well-known neurological disease, Parkinsonism, which a neuropathologist, Oleg Hornykiewicz[13] had found was due to a drastic reduction of dopamine in certain areas of the brain. This led to the treatment of Parkinsonism with the dopamine precursor, L-Dopa, but large doses often caused patients to suffer from disconnected thoughts and other symptoms of schizophrenia. These side-effects caused by drugs, demonstrated a possible relationship between neurological and psychiatric disorders, and also between alterations in behaviour and molecular change.

On the basis of all these observations, made in diverse areas of pharmacology and medicine, in 1967 a catecholamine hypothesis was formulated by Joseph Schildkraut.[14] The basic suggestion was that mania, whose symptoms are a state of excessive mental and physical activity and elation accompanied by flights of fancy, was perhaps due to an increase in the availability of the two catecholamines, dopamine and norepinephrine in certain regions of the brain. It was also suggested that depression was caused by a depletion of these amines, which might be the cause of the feelings of sadness, lethargy and hopelessness. As interest in this simple hypothesis spread, it was extended to include the schizophrenias – a very complex group of diseases characterised by thought disorders, disorganised speech, hallucinations, and delusions. The new hypothesis suggested that these disorders were caused by 'chemical imbalance' in the brain, involving these biogenic amine transmitters. This could also account for the effect of psychoactive drugs and for the analytical results on metabolites.

Finally, during the peak of the college 'drug scene' in the 1960s, the difficulties in distinguishing between hallucinatory drug-induced psychoses and schizophrenia suggested that the effect of such drugs like LSD might be due to the overall similarity of their chemical structure to that of the amine neuro-transmitters. A new hypothesis evolved: that tranquilising and other drugs may act by competing with or displacing neuro-transmitters. In order to examine and evaluate the biological approach to mental disorders a new discipline, Biological Psychiatry, was born and established in many medical schools and mental health research laboratories. The birth of this new area of endeavour was accompanied by the publication of special journals, the foundation of specialised societies and a claim for a special place in the curriculum. These are all the indications of a 'paradigm' in the terminology of Thomas Kuhn,[15] 'of achievement sufficiently unprecedented to attract adherents away from competing modes of scientific activity, and sufficiently open-ended to leave all sorts of problems for the redefined group of practitioners to resolve.' And there *are* many problems for these practitioners to resolve, even in planning experiments for testing the basic hypotheses. First, the difficulty of reliable psychiatric diagnosis – at one time the classification of mental disorders, was considered to be almost an academic exercise, but it now becomes an important prerequisite not only for the correct choice of treatment, but also as a basic requirement for further research.

Psychiatric diagnosis is not yet an exact science and there are, in fact, no absolute measures. In many cases reliable data can only be obtained by more than one opinion, by correlations between observers. Then, studies on human patients, even volunteers, are limited both by the rules of medical ethics and, as in the case of memory research, by the inaccessibility of the brain to anatomical or chemical manipulations. In the study of memory and learning, and even of cell recognition, model systems can be used. Even simple invertebrates, such as fruit flies, can be taught to modify their behaviour and retain their memories for a reasonable length of time. However, no useful animal models of mental disorders were known. It appears, from studies in zoology and ethology, that few if any aberrant animals exist or could survive in the wild. Considerable effort and ingenuity have therefore been expended to produce 'abnormal behaviour' in experimental animals – such as 'psychoses' in monkeys (caused by isolation) and 'depression' in rats (caused by reserpine or electric shocks).

Human beings are obviously much more complex, with highly developed nervous systems and greater sensitivity to stress, and no animal models can unequivocally reproduce the human situation. Even manipulations of the brains of animals by drugs or surgery have not been entirely successful: injections of amphetamines induce 'psychotic' behaviour in rodents and LSD may cause cats to hallucinate, but many basic problems remain; animals do not seem to have the capability of symbolic thought and are probably less prone to the detachment of their mental processes from reality.

Nevertheless, using animals, a great deal has by now been clarified about the basic chemistry and enzymology of the brain and many new drugs have been tested and used successfully for therapy. The amine hypothesis has given enormous impetus, direction and purpose to research, though one should not forget that over-emphasising and reiterating hypotheses based purely on research with animal models may lead to over-interpretation and over-speculation. Ultimately it is obvious that the only rigorous test of any hypothesis of mental disorders must be carried out in patients themselves. Here ethical and practical problems must certainly be considered, and a safe method found for determining significant chemical changes in the brains of human subjects.

A well-tried technique for investigating biochemical processes in general is to attach a label to a molecule so that it can be identified, or measured at any time or place in a biological system. This is done by substituting some atom in a molecule by its isotope, which does not affect the chemistry but provides the label. In biochemical and pharmacological research, including much of the work on animal models of psychoses, radioactive isotopes have been used as labels – usually carbon-14 or tritium (radioactive hydrogen). These emit ionizing radiation by means of which they can be detected. But the use of radioisotopes as tracers in humans is no longer permitted and when alternative methods of 'looking inside the head' were tried, it was found that the brain is protected from outside intervention, not only by the skull, but also by a series of membranes and biological barriers which are permeable only to essential materials. Even dopamine cannot get into the brain from outside and for this reason its precursor L-Dopa is used to treat Parkinsonism. One solution, for biomedical research in human subjects that has been suggested is to use stable (non-radioactive) isotopes for labelling. These do not emit radiation, but can be detected by their weight. Such isotopes are known for oxygen, as I mentioned earlier, and can be separated and concentrated in oxygen gas, which is fortunately freely accessible to all regions of the brain. There some of it becomes part of the biogenic amines that we have been discussing.

Heavy oxygen or oxygen-18, which is produced by the Weizmann Institute, was proved to have no harmful effects whatsoever on animals even after very long periods of breathing it. It was also shown that organic materials, including amine metabolites containing oxygen-18, can be detected with great precision by physico-chemical analytical techniques, many of which have been developed in Sweden. During the past four years, together with Goran Sedvall, a

Swedish psychiatrist and professor of pharmacology at the Karolinska Institute in Stockholm, we have examined the turnover of biogenic amines in the central nervous system, first of rats and monkeys,[16] and more recently with volunteer patients. In a typical experiment the subject breathes air containing heavy oxygen (using a simple breathing apparatus) which is transported to the brain to become part of the amines. The label is found later, in a characteristic pattern in the metabolites of neuro-transmitters isolated from body fluids. In this way we have labelled the molecules that may cause madness. Next, the effect of anti-psychotic drugs, such as haloperidol, has been examined in volunteer neurological patients. There already seems to be evidence of competition between drugs and transmitters – though this does not yet validate the whole amine hypothesis. There are still many problems to be solved, variations in metabolism between one patient and another, often in the same patient himself, due to personal stress or changes in hormonal level. Another, more serious difficulty is that these chemical effects are probably regional, highly localised in one part of the central nervous system and similar chemical processes take place elsewhere in the brain and body. We are now trying to solve this by returning to more detailed regional studies in animals. The results, together with similar studies from many other laboratories, will, I am sure, modify the original simple hypothesis. The new data will help determine the time scales with which behavioural changes and biochemical effects occur and how drugs affect them – thereby clarifying what is cause and what is effect.

In the past year there have already been suggestions that anti-anxiety agents affect other neuro-transmitters, both excitatory and inhibitory, which therefore also have a role to play in the more subtle changes of mood. The range and variety of symptoms of major mental disorders may indeed depend on the interaction of many chemical pathways in the brain.

It is also becoming evident that heredity plays an important role in some of these diseases and also that the environment in its broadest sense cannot be ignored. Schizophrenia is often said to run in families, but a family shares not only its genetic endowment, but also its environment and either of these could account for such familial tendencies. Seymour Kety[17] and his colleagues have studied the prevalence and type of mental illness of people adopted as children in Denmark. These children carry a genetic heritage from their biological family but live in the environment provided by an adoptive family, and the statistics seem to support the operation of genetic factors in the transmission of at least one form of schizophrenia. These results are still not universally accepted although great care was taken in the diagnosis and in the statistical treatment of the results.

I have deliberately emphasised the state of the art in two rather extreme instances of current scientific investigation. I have chosen these aspects of brain research not only because I am fairly familiar with their background and current trends, but mainly to demonstrate by specific examples the basic problems of research in general – and research on the frontiers of science in particular.

Above all else scientific enquiry has two major goals: to understand and to apply, its subject matter. Research into memory is aimed at the first goal – an effort to understand the complexities of what goes on in our own heads, and to bridge the gap between disciplines with long-established methodologies (such as chemistry and physics) with psychology, which is still in the process of looking for clear and systematic ways of describing the behaviour of living creatures, especially that of human beings. For this reason maybe memory research is a field to which many scientists are drawn, out of sheer intellectual curiosity; and is a model of science done for its own sake – in which immediate practical applications are few and even these not always self-evident. Research into memory suffers from an overabundance of data scattered throughout a wide range of disciplines, methods and approaches. It needs great patience to find and sift out salient points on which to build a coherent set of relationships and correlation. Finally, there is also an area in which investigators must know when to persevere and when to stop, for while there is no dearth of partial hypotheses, there is still no overall picture. Like so many other areas of fundamental research, progress is slow; we still await the man or woman of heightened intuition, who will eventually reveal what remains hidden and provide us with an explanation of what is not understood.

By constrast, the other area of research I have described concerns itself with the second goal of science – its application. It is aimed at bridging the gap between our knowledge of the functioning of the brain and abnormal behaviour. Interest in *this* research is promoted largely by the search for solutions to human distress.

I need not, I think, elaborate on the urgency of the overall problem of mental diseases; its magnitude is equalled only by our ignorance. Although no scientist knows yet how people can be made happier, there are those who strive, at least, to lessen their unhappiness. Scientists in laboratories are not immune to social pressure, particularly when the problems are clearly spelled out. The trend towards mission-oriented research has been effectively encouraged by governments, by funding agencies, by the media and by the public at large. In other fields such as agriculture, medicine and the various forms of engineering and technology, the application of science has indeed resulted in practical solutions. But this kind of research is often undertaken under the pressure of timetables – and thus is inevitably less than complete – something which may only become apparent when it is almost too late for remedy. Sadly enough, the promising fruits of applied science are found and not rarely, to have hidden worms. The resultant disappointments and increase in public scepticism lead in turn to the questioning of the basic value of the scientific enterprise. This is partly due to a lack of knowledge (or is it understanding?) of the complexities of science and, I fear also, to some loss of confidence in the scientists themselves. The answer, I suggest, may be a return to basic methods; to the cautious and self-critical approach, taking time to evaluate relevant facts and consider each and every possible outcome.

Science, by its nature, needs to be unhurried, even undirected sometimes; scientists must remember that though they should be audacious in seeking hypotheses, they must be meticulous in experimentation. By returning to the initial quest, to the understanding of basic principles – such as the workings of the normal waking brain – we will also, I am sure, in due course find solutions to the dire problems of malfunction – to stress, to over-anxiety, to aggression and to psychoses. Let me end by borrowing the words of Xenophanes:[18]

> The Gods did not reveal from the beginning
> All things to us, but in the course of time
> Through seeking we may learn and know things best . . .

11th March, 1977

References

1. C.G. Hempel, Fundamentals of Concept Formation in Empirical Science. *International Encyclopaedia of Unified Science*, vol. 2, No. 7, p. 1. University of Chicago Press. (1952).
2. K.R. Popper, Science: Problems, Aims, Responsibilities. *Federation Proceedings*, vol. 22, p. 961. (1963).
3. R. Robinson, So-called Anomalous Water, *Chemistry and Industry*, p. 782. (1969).
4. F. Bacon, *Novum Organum*, vol. VIII of *The Works of Francis Bacon* ed. J. Spedding, p. 210. R.L. Ellis and D.D. Heath; New York. (1869).
5. H.C. Urey, G.B. Pegram and J.R. Huffman, Concentration of the Oxygen Isotopes. *Journal of Chemical Physics*, **4**, p. 623. (1963).
6. A. Einstein, Preface in: *Where is Science Going?* Max-Planck, translated by J. Murphy. George Allen and Unwin Ltd: London. (1933).
7. H. Ebbinghaus, *Memory, a Contribution to Experimental Psychology*. (First published in 1885) translated by H.A. Ruger and C.E. Bussenin, Dover Publications Inc. (1964).

8. D.O. Hebb, *The Organisation of Behaviour*, p. 1. John Wiley & Sons Inc. (1949).
9. K.S. Lashley, In Search of the Engram: Physiological Mechanisms in Animal Behaviour. *Symposia of the Society for Experimental Biology*, No. 4. (1950).
10. J. Von Neumann, The Computer and the Brain (1958), quoted in D.E. Wooldridge, *The Machinery of the Brain* p. 188. McGraw-Hill Book Co. Ltd. (1963).
11. E. Tanzi and I. Fattie, Le Induzioni nell'odierna Isotologia del Sistema Nervosa. *Revista Sperimentale di Freniatria*, **19**, p. 419. (1893).
12. J.H. Quastel, Biochemistry and Mental Disorders. *Lancet 11*, 1417. (1932).
13. O. Hornykiewicz, Dopamine and Brain Function. *Pharmacological Reviews,* **18**, p. 925. (1966).
14. J.J. Schildkraut, The Catecholamine Hypothesis of Affective Disorders: A Review of Supporting Evidence *American Journal of Psychiatry*, **122**, p. 509. (1965).
15. T.S. Kuhn, The Structure of Scientific Revolutions. In the *International Encyclopaedia of Unified Science*, Vol. 2, No. 2, University of Chicago Press. (1962).
16. D. Samuel, The Measurement of Biogenic Amine Turnover Using Oxygen-18. In *The Impact of Biology on Modern Psychiatry*, edited by E.S. Gershon, R.H. Belmaker, S.S. Kety and M. Rosenbaum, p. 95, Plenum Press: New York. (1977).
17. S.S. Kety, Genetic Aspects of Schizophrenia: Observations on the Biological and Adoptive Relatives of Adoptees who Became Schizophrenic. Ibid, p. 195. (1977).
18. Quoted in *Popper* by Bryan Magee, p. 28, Fontana/Collins: London. (1973).

LECTURE VI

Nature in a Mirror

Glynne Wickham

Predictably – given the title I have chosen for this lecture – I start with Hamlet's advice to the players.

> Be not too tame neither, but let your own discretion be your tutor; suit the action to the word, the word to the action; with this special observance, that you o'erstep not the modesty of nature; for anything so overdone is from the purpose of playing, whose end, both at the first and now, was and is, to hold, as 'twere, the mirror up to nature; to show virtue her own feature, scorn her own image, and the very age and body of the time his form and pressure.
>
> (Arden ed., III ii 19–29).

From the moment Richard Burbage first uttered those words in 1601 or 1602, the belief has grown that Shakespeare was here using Hamlet as his mouthpiece to make a personal statement about the art of acting, and that what Shakespeare was advocating was the cultivation of a naturalistic or behaviourist technique as the prime objective of an actor's art and training. The fact that, like Polonius, Hamlet himself was an amateur actor – patently the President-elect of W.U.D.S., or Wittenberg University Dramatic Society, had he been allowed to stay there – has been almost totally ignored, Harley Granville-Barker nobly excepted: so too has the fact that the advice itself sounds much more like what we might have expected to hear from Sir Philip Sidney, or Ben Jonson, than from Shakespeare. Either way, however, a moment's reflection reveals that the advice is ambiguous. Indeed its very ambiguity is perhaps the most important aspect of it, since this reflects so accurately the great argument that engaged the attention of everyone concerned with dramatic art in the Elizabethan and Jacobean era about the moral purpose of stage-plays and the legitimacy of acting as a profession.

Shakespeare, of course, was himself deeply involved in this controversy on both counts; but in general he chose to steer clear of it and, whatever he may have said about it in talking to his friends, he avoided committing his views to print.

No one today is troubled about the legitimacy of acting as a profession, although the green-eyed monster of jealousy still raises its shrill voice from time to time in protest over some aspect of the life-style of film stars and pop-idols, or about financial rewards that appear to bear no relation to what most citizens understand as 'a full day's work'. Other voices occasionally claim that TV plays pollute our homes and damage children's minds with gratuitous bombardment of fortuitous violence and obscenity. In Shakespeare's lifetime, however, not only were actors scorned and abused by many learned and wealthy men on these same counts, but they had to live with a constant threat to their livelihood from a section of the community that sought to ban all plays and acting in any well-ordered State on moral grounds. That this threat was not an idle one may be judged from the simple fact that, in 1642, when Charles I and his Court had left London for Oxford, Parliament closed all our Theatres by Statute *sine die*. By the same token all actors bar one, Hilliard Swanston, joined the King's army in the Civil War; and, when it was over and acting was again permitted following the Restoration, Sir William D'Avenant observed, 'They that will have no King, will have no play!'

Why, then, should plays and acting have stirred such deep feelings 400 years ago? And how is it that Jean Giraudoux, France's most sophisticated playwright of the interwar years could state, as he did in the 1930s, '*Le spectacle est la seule forme d'éducation morale ou artistique d'une nation*'?

Where then do we stand in regard to dramatic art? Is it just frivolous ephemera, a harmless relaxation for lazy-minded people at the end of a day's work as those millions who never enter a theatre's portals (except for the Christmas pantomime) suggest by their studied neglect of it? Is it a dangerous and insidious pollutant of mind and soul, a 'Market of Bawdry' as Stephen Gosson described it in Shakespeare's lifetime, and an invitation to damnation as John Knox proclaimed? Is its existence a blot on the escutcheon of an ideal Republic as both Plato and Calvin maintained? Or is it something so worthy of respect as to justify both Giraudoux's apostrophe of its educational virtue and our own Arts Council's lavish expenditure of tax-payers' money on its physical support? Put very simply, do we set any real store by the possession of a National Theatre, or a British Theatre Museum, or provincial repertory companies, or departments of drama in our universities, all of which, as tax-payers, we are obliged to subsidise. Or, if we are honest with ourselves, would we rather be without them, and see the sums of money presently devoted to keeping them afloat, diverted to some other purpose? I say 'diverted' since no sane person can imagine a Government today, of whatever political complexion, returning the money to our own pockets! Let us not deceive ourselves: the entire fabric of our national and provincial theatrical enterprises would collapse overnight like a house of cards if the subsidies that alone make them viable were once to be withdrawn.

As production costs soar with inflation, and actors' and technicians' wages rocket proportionately, ever larger subsidies are required merely to preserve the *status quo*. Sooner or later, therefore, the questions that all of us will have to answer are where and when they must stop. That I am not taking recourse to doom-mongering in saying this is borne out by the newspaper poster which greeted me on arrival at Paddington today – 'National Theatre going bankrupt'. Curiously, if these same questions were to be asked in the form of one of Mrs Thatcher's promised 'referenda', I suspect the answer, in terms of theatre, would be a resounding 'no' to subsidy – between 5% and 10% for, and 90% to 95% against, since statistically those are the known percentages of our population that, nationwide, respectively do and do not buy theatre tickets at the box office. If the same question were raised in terms of television I suspect that this position would be reversed with perhaps 30% against and 70% in favour. If that is true, is the live theatre, as we know it, doomed to be subsumed by television by the end of this century? No one, as yet can be sure of that; but the mere possibility is surely enough to invite thought and discussion of what we really want to see on our magic boxes, our silver screens and in those archaically uneconomic buildings which we still call theatres? Or do we want drama in our time to gravitate to other, simpler, environments – pubs, clubs, streets and village halls – and to be characterised by a more spontaneous, improvisational style?

I will approach this topic by returning to Hamlet and his remarks on the actor's art. He insists, you will recall, that the players shall not,

> o'erstep the modesty of nature

and that the whole 'purpose of playing . . . was and is, to hold, as t'were the mirror up to nature.' But what sort of mirror? Reflecting? Distorting? One way? Flat? Convex or concave? He doesn't say. Instead, he claims that this particular mirror must,

> show virtue her own figure, scorn her own image, and the very age and body of the time his form and pressure.
> (*pressure*: lit.; impress, device)

How do you do this? Surely, that is a stiff challenge to any normal mirror known to you or me?

A TV camera, taken out into the local High Street effectively mirrors what is passing: cars (some large and glossy, some small, others scratched or dented); pedestrians (some male, some female, some in twos or threes, others alone, some brisk and jaunty, others with a hunted look). In short, it mirrors the passing scene, but hardly 'virtue her own figure' or 'the form and pressure of the time'. And indeed precisely because of this, between the cameraman and we, the viewers, stands the interviewer, the vision mixer and the director, all of whom are busy interpreting the scene, editing it, and thus determining what we shall and shall not see; what we shall and shall not believe. It is through these processes, discreet or indiscreet, that virtue is differentiated from vice (alternatively, deliberately left undistinguished) and that we are conditioned to receive or to ignore the distinction. Commercials supply the most self-evident example, since our natural credulity is there being stretched as far as the advertiser thinks he can safely stretch it. Yet the virtues of most soap-powders, as all of us know well, were carried far beyond that limit long ago!

If men and women of the first Elizabethan era were not technically equipped to recognise this particular distorting property of the dramatic mirror, they were certainly more acutely aware than we are of another of its distorting faces – its subversive power in matters political and religious. Precisely on this account, a succession of Tudor Governments from 1543 onwards felt obliged to ask Parliament to legislate against the theatre imposing the restrictive reins of censorship upon it in an endeavour to tame it and, later, to destroy it. As these pressures mounted, so the counterweight of royal prerogative had to be used to defend it. James I took the decisive step in 1603 of taking the three leading acting companies into his own, Queen Anne's and Prince Henry's personal Households and banning all others from acting in London. We are apt to forget that, but for this action, it is highly unlikely that any of Shakespeare's plays from *Measure for Measure* and *King Lear* to *Henry VIII* and *The Tempest* would have been written. Even so, James himself was forced to intervene on several occasions to remind both actors and playwrights that continuance of royal license to present plays in public depended on their willingness to keep the subject-matter of their plays within the bounds of orthodox opinion. Breaches of this understanding nearly cost Ben Jonson his life in 1605, made it possible for the Burbage/Shakespeare company to recover their lost lease of the Blackfriars playhouse in 1609, when Evan's boys were banned from playing there, and got them arrested in 1623 for insulting the Spanish Ambassador. When Hamlet, therefore, proposes to present a play at Elsinore, and when King Claudius says to his Lord Chamberlain, Polonius, 'Is there no offence in't?' his question is neither an idle one on his part, nor just a convenient fantasy of Shakespeare's own coining. The question is itself a lively mirror-image of the times.

Two questions arise from this state of affairs. How was it that the Elizabethan and Jacobean theatre came thus to be deemed so dangerous to the interests of the State as to warrant the imposition of such tough controls upon it? And how, in the first instance, did it ever become so deeply entangled with politics as to provoke this rough response? To answer those questions, we must retreat still further into time and look briefly at the religious drama of the Middle Ages: for there we come face to face with an explicit declaration of intent. In a world governed by the still single and undivided Church of Rome, drama was conceived of as an instrument of education. It existed to celebrate festive occasions and to comment on them. Above all, it existed to explain the significance of the occasion to largely illiterate audiences: in short, in Hamlet's words 'to show . . . the very age and body of the time his form and pressure.' It is for this reason that the great English Cycles of Mystery Plays took the name of the annual Festival they were invented to celebrate – 'The Plaie Called Corpus Christi': and it was for similar reasons that virtually every town in the Kingdom possessed a Saint's Play rehearsing the life and works of the patron Saint of the parish, or of the guild, on that day in the year particular to the Saint in question. Frequently this occasion was used to raise money for the fabric of the church, or the relief of the poor in the parish. Thus the performance was at once devotional, instructional, and strictly practical in its purpose.

In every case it was the abiding significance of the life or events celebrated, rather than the events themselves, that medieval actors and playmakers sought to impress upon their audiences. Biblical characters were thus depicted neither as historical persons from another epoch, nor in geographically correct replicas of their historical settings, but as contemporaries of the audience, and in recognisably local costumes and settings. The High Priests, Caiaphas and Annas, therefore appear in the Mystery Cycles as an English Cardinal and an English bishop; Pontius Pilate becomes Sir Pilate, JP; Pharisees and Saducees are depicted as laywers from Chancery; Bethlehem shepherds are translated into shepherds of the Pennines or Cotswolds; Mary Magdalene becomes a London prostitute, and so on. Dialogue is frequently anachronistic, not because the authors knew no better, but because they wished to impress upon their audiences that treachery did not die with Judas Iscariot, that hypocrisy was not unique to Pharisees, and that tyranny and vested interest were as prevalent in the feudal society of fifteenth century England as in the Judea of Herod Aggripa, and would remain so throughout time until Judgement Day itself.

Inevitably, such plays contained much social criticism: and as many of them – more especially those known as Moralities – set out to distinguish Virtue from Vice and to urge repentance upon sinful mankind while time availed, this criticism was grounded on the bedrock of those moral and ethical values that distinguished a Christian way of life from a pagan one. These distinctions thus became the primary images reflected in the mirror of medieval plays.

Granted an overtly didactic, community drama as popular and widespread as this had become by the sixteenth century, the advent of the Reformation could only serve to drag it into the political arena, since the doctrines which the plays were designed to illustrate and establish were those which the State now sought to stamp out. At the same time, it was realised that with only a few deft changes here and there, such plays could be translated into weapons of propaganda to advance the Reformation. 'For Vices and Devils' says the playmaker to his audience, 'read Romish priests and bishops; for Anti-Christ read the Pope! For the Virtues you may substitute us Reformers.' This was the technique adopted by Sir David Lyndsay in *Ane Satyre of the Thrie Estaits* and by Bishop John Bale in *Kyng Johan* in the 1530s. So vividly did they and others mirror the major religious and political controversies of their day upon the stage, that the mere reflection of these antipathies provoked rioting inside and outside the auditoria. Police action followed as Henry VIII endeavoured to control this unexpected and unfamiliar situation; and as is the way with such actions, censorship and persecution followed in their wake. Yet it is one thing to forbid explicit reference to living persons and specific opinions by recourse to legislation; quite another to prevent playmakers who are already accustomed to using allegory from continuing to do so and thus effectively circumventing the legislation. In short, within this tradition it remained easy to say one thing and mean another. Thus playwrights quickly learned to distance their subject matter into the historical past, and having done that, so to depart from historical fact as to supply, in fiction, close parallels to contemporary events. Continued use of abstract personifications, rather than specific individuals served only to create a yet thicker smoke-screen to protect the author and his actors from charges of heresy and treason. Yet a few, well-chosen allusions and associations scattered here and there in the dialogue, or in stage directions, suffice to enable audiences to link the two together for themselves. This may seem to us to be too complicated a game to play with enjoyment; yet it is neither more nor less difficult or enjoyable than learning to identify aircraft by their silhouettes or an engine by its sounds. If the playwright is challenged he has only to disclaim any intentional *double-entendre* and attribute its existence to the imagination of the spectator.

It was by these means that the anonymous author of *Jacob and Esau, c.* 1550 while seeming innocuously to retell an already familiar Old Testament story in stage-terms, in fact argued that Roman Catholics had sacrificed their birthright in England to the Protestant Reformers who, like Jacob before them, represent God's 'elect'. Thus not only does this play become a mirror of Reformation polemic; but, still more daringly considering its date, it introduces the Calvinist doctrine of predestination to English audiences. Still more significantly, and dangerously, it advances the idea that God's will, as revealed by inspiration, is above and beyond that of Kings and governments – an idea that was to lead, in the fullness of time, to the execution of Charles I. We could do worse than recall today that it also underpins the philosophy of terrorism. Be that as it may, it was by such means that Marlowe, in *Dr Faustus*, despite the Elizabethan ban on all overtly religious plays, could explore with his audiences the predicament of the Calvinist intellectual who knows himself to be numbered among the reprobates, yet who lacks adequate faith in God, or in his promised mercy, to repent. And it was by these means that Shakespeare could contrast in *King John* the morality with the expediency of the execution of Mary Queen of Scots, or comment on the Gunpowder Plot in *Macbeth*.

Awareness of these techniques, therefore, and skill in the use of them, enabled English playmakers of the Elizabethan and Jacobean era to employ the art of drama in the way in which Hamlet urges on the players from Wittenberg, 'to show virtue her own feature, scorn her own image, and the very age and body of the time his form and purpose'. What, however, these techniques paradoxically forbid is photographic realism in the representation of the events or persons mirrored of the kind we have come to expect from the cinema. On the contrary, they oblige the playmaker to employ emblems, symbols, allusions, and correspondences that will serve simultaneously to stimulate imaginative perception in the auditorium and separate mask from face, outward semblance from inner reality, play from earnest. In other words such a play is a special form of game: a game employing a visual and a verbal language of signs and figures, carefully arranged by the playmaker and executed by the actors so as to give concrete, outward form to abstract concepts, thus transcending the normal frontiers of literacy, and communicating complex, religious, political, social and psychological ideas in a swift, but ordered succession of active, three-dimensional images. In such a theatre – a stage that caters as readily for the abstract concepts of heaven and hell as for the materialistic, natural world – kings must rub shoulders with peasants, merchants and their wives, with farmers and courtiers, princes with gravediggers and beggars, generals with common soldiers. There can be no artificial separation of persons of high degree from those of humbler birth if the stage-mirror is truly to reflect nature. Nor must it be pre-supposed that, to be a tragic hero, a man or woman must necessarily be good, nor that comedy must exclude all reference to painful situations or vindictive people. Time, moreover, is an essential ingredient in effecting reconciliations, in snatching joy from despair, or in unmasking the seemer who smiles, and smiles, but is a villain.

Thus if Shakespearean drama constantly flouts or ignores classical insistence upon the unities of time and place and action, it is because it sought – however misguidedly – like its medieval antecedents, to mirror the universe as understood within Christian theology; to reflect the macrocosmos in the microcosomos of *Theatrum Mundi* – 'the great stage of the world' where all men and women are merely players acting out their lives before the Almighty, their Creator, Saviour and Judge.

The tradition stemmed ultimately from St Augustine and the notion that every visible and tangible object in this world was no more than a pointer, a signpost towards the ultimate reality in the City of God. All such signs were deceptive and unreliable; but, with care and discretion, they could be used to assist discovery of the nature and purpose of mortal existence. Likewise, stage plays, being fictions, – imitated actions rather than real ones – are simply signs; as is clear when all the corpses that litter the stage at the end of *Hamlet* rise to take their curtain call and repeat the peformance tomorrow. They are pointers to truth or reality, rather than truth or reality itself. The art lies in penetrating the plethora of outward semblances in order to mirror accurately what lies behind or beneath them, not in copying them, or classifying them, or codifying them. Yet all these stage-actions must be so ordered by the playmaker, and so executed by the actors as not to stretch our credulity as spectators further than we are willing to let it go: to that extent they must appear 'natural': and therein lies the force of Hamlet's 'special observation' to the players – 'that you o'erstep not the modesty of nature'.

It is in his ensuing words, however, that ambiguity resides; the comment that the purpose of playing 'both at the first and now, was and is, to hold, as 'twere, the mirror up to nature'. Even some of Shakespeare's own contemporaries, as he himself well knew, disagreed. Chief among them was Sir Philip Sidney whose comments on the theatre in *The Defense of Poesie* had been published some three years before *Hamlet* was written. No less vocal on the point was Shakespeare's younger colleague, the actor-playmaker Ben Jonson. Both insisted that a play should not be a mirror-image of life, but an argument about life. Effectively, what they and their followers sought to do was to substitute a rational and objective approach to dramatic art for the subjective and irrational one which, in their view, had governed it for the past five hundred years. Adopting a typical renaissance stance, they sought to short-circuit the traditions of the Christian Gothic past and return to principles articulated by the critics and theoreticians of clasical antiquity: and where they lacked adequate knowledge of Greek opinion they were content to accept Roman pronouncements. Tragi-comedy was the prime target of their attack, a mongrel product not countenanced (as they thought) by either the Greeks or the Romans. But as most of life as we know it is in fact a constant mixture of joy and sorrow, and only quite abnormally an unmitigated sequence of either triumphs or disasters, their assault on theatrical tragi-comedy was in fact an attack on the very idea of holding 'a mirror up to nature'.

What, then, did they want instead?

First, they wished to re-establish what they took to be clear-cut classical distinctions of dramatic genre. Tragedy, to be worthy of the name, must be serious. Basing their case on the writings of Aristotle, Sophocles, Seneca and Horace they hoped to re-establish classical principles of play-construction, banishing comic incident and characters like those associated with the grave-diggers in *Hamlet*, the porter in *Macbeth* or the fool in *King Lear*, and confining the cast-lists of tragedy to kings and courtiers. Deaths must take place off-stage and be reported. The entire stage-action must be restricted to a single locality and concentrated within a twenty-four hour time-span. By the same token, comedy must confine its cast-lists to commoners, and thus to bourgeois concerns of the market-place and domestic life.

Secondly, the *avant-garde* among Jacobean playwrights and critics sought, by codifying drama in this manner, to rationalise its moral purpose and thus to shift the core of its appeal from raw, emotional responses to a refined, intellectual appreciation. The serious arguments of tragedy must prepare its recipients to accept life's calamities and injustices with stoic resignation: the witty arguments of comedy must expose human credulity and then ridicule such folly. The shift, in comedy, is thus away from the romantic concept of love, mercy and forgiveness as agents of reconcilement in society, and towards a more realistic ethic that is at once cynical and punitive: the shift in tragedy is oriented away from the controls imposed upon mankind by an all-seeing, paternalistic and avenging Divine Providence towards acceptance of inscrutable laws of chance and natural selection. Thus by the end of the seventeenth century, in accordance with the scientific spirit of the age, tragedy no longer seeks to depict the release of violent passions and their consequences, but starts with all passion spent, and then proceeds to analyse the spending of it in the light of the possible alternatives: comedy by contrast, simply depicts manners and urges men, if they are wise, to restrain their instincts and avoid entanglements. Drama, in this condition, can thus still be said to hold a mirror up to nature, but the nature of the mirror itself has changed. No longer are we invited to look at life as it should and might be; rather does this ask us to view life as it was actually lived by a small section of society – courtiers and their acquaintances – and then draw our own conclusions. Thus, where the former mirror is angled to present a broadly optimistic view of life, the latter is angled to reflect a pessimistic one.

Once it is realised that it is possible so to manipulate the mirror images of life presented in dramatic art as to accommodate both viewpoints, it becomes easier to understand how the case arises for banishing this art-form altogether from the Utopian State. Plato espoused this cause in Periclean Athens; so did John Calvin in sixteenth-century Geneva; but only in the Islamic world and in Britain has it actually been put into effect. It was the Islamic invasion of India that destroyed the ancient Sanscrit drama which, ironically, had to await the arrival of the British Raj to re-emerge from its long sleep. In these islands we have to thank John Knox and Lord Protector Cromwell for this dubious distinction.

The twenty year ban on acting in Britain stemmed from four principal sources, one of them being the interest of renaissance humanists in the writings of Plato and Aristotle. Another was Jewish. The other two were specifically Protestant. The Jewish factor was those Old Testament verses forbidding tranvestism. The Elizabethan preacher, Stephen Gosson, summed up the matter succinctly.

> The Law of God very straightly forbids men to put on women's garments.
>
> (*Playes Confuted in five Actions*, 1582).

Thus, in the wake of the Reformation, although the wearing of surplices by priests in Britain's churches aroused as much passion as the sight of boys playing female roles on British stages, it was the theatre at large that took most of the blame. Gosson continued

> In Stage Playes for a boy to put on the attyre, the gesture, the passions of a woman; for a meane person to take upon him the title of a Prince with counterfeit porte, and traine, is by outwarde signes to shewe them selves otherwise than they are; and so within the compasse of a lye.
>
> (*Ibid.*)

Another preacher, Philip Stubbs, went further, describing London's two, public playhouses as 'Venus' palace and Satan's synagogue'.

> Doo these Mockers and Flowters of (God's) Majesty, these dissembling *Hypocrites*, and flattering *Gnatoes*, think to escape unpunished? beware, therefore, you masking Players, you painted sepulchres, you double-dealing ambidexters, be warned betymes, and, like good computistes, cast your accounts before, what will be the reward in the end, least God destroy you in his wrath.
>
> (*The Anatomie of Abuses*, 1583).

The first of the specifically Protestant attacks on plays derived from an equation of theatre-going with idleness and extravagance, the deadliest vices in the canon of the new faith, as opposed to the two cardinal virtues of work and thrift – the corner-stones of emergent, capitalism. The second was the continuing association of plays and play-acting with Roman Catholicism. As William Crashaw remarked in a sermon at Paul's Cross in 1608,

> 'The ungodly plays and interludes so rife in this nation' – he is here speaking of Shakespeare, Jonson, Dekker, Webster,

> *et al*, we might note – 'what are they but a bastard of Babylon, a daughter of error and confusion, a hellish device (the devil's own recreation to mock at holy things) by him delivered to the Heathen, from them to the Papists, *and from them to us*'.

This last thrust was as near to being mortal as any that could be imagined: for notwithstanding the relaxation of the ban which followed the accession of Charles II, drama continued to be regarded in most middle class homes throughout the country as aligned with Popery and to be shunned like the devil – especially so in schools and universities – until the start of this century.

It is to be doubted whether this would have been the case had the mirror of Restoration tragedy not reflected so pale and esoteric an image of nature in its interminable discussions of love in conflict with honour (in rhymed couplets at that), and had the image of society reflected in the witty arguments of Restoration comedy not been so offensively critical of merchant-bourgeois morality as to provoke Jeremy Collier in 1698 to launch an attack on what he called 'The Immorality and Profaneness of the English Stage'. To these fires were added those provoked early in the eighteenth century by Henry Fielding's and John Gay's flagrantly censorious criticism of cant and hypocrisy among professional men – lawyers, doctors, the clergy and, above all, politicians. Not since the days of Aristophanes had anything that was at once so outspoken and so amusing, been seen or heard in the theatre. Sir Robert Walpole, sensing himself to be publicly ridiculed, retaliated with the Licensing Act of 1737 which, for nearly two centuries, ruled out any serious discussion of politics, religion or sex on the English stage. Posterity may thank Walpole for thus deflecting Fielding's energies towards the writing of novels; but it has him to blame for enabling the poet Humbert Wolfe two hundred years later to remark in his *ABC of the Theatre*:

> C is the censor, whose duty has been
> To rule out God, and the King, as obscene.

Glynne Wickham

Who remembers today that Nugent Monk, founder of the Maddermarket Theatre in Norwich, was arrested in London in 1909 on the instructions of the Lord Chamberlain for staging the Passion sequence of *Ludus Coventriae*? Indeed, it was not until 1968 that the last of these impediments to freedom of speech in the theatre was finally jettisoned.

It is hardly surprising, therefore, that so few English plays written in the eighteenth and nineteenth centuries have much of importance to say about life as it was lived then, or can be deemed to possess much literary merit now. But are these adequate grounds for dismissing the Victorian theatre as frivolous and vacuous? Or, if we are tempted to do so, are we sure we are looking in the right mirror? Did not Victorian drama, within the limits imposed upon it by the censorship, aim to reflect the industrial revolution and the twin theatrical phenomena that accompanied it – the influx of proletarian audiences, quite untrained in the literary conventions of dramatic art; unable either to concentrate on argument for long or to appreciate niceties of language; and who demanded stories, song, dance and spectacle – coupled with rapid growth in theatrical technology to support this demand for strong narrative and spectacle? These pressures were met by new and enterprising managers in Music-Halls, Circuses, Hippodromes and Palaces throughout the country where the fears and fervour generated by the French and American revolutions were fuelled with goulish sensations and lurid romance pillaged from Gothic novels and penny-dreadfuls. The substitution of gas-light for candles opened the way to the darkening of the auditorium and the novelty of lighting effects on the stage that could be controlled manually during the performance.

A particularly virulent ingredient of this new mixture was the rage for drama with an equine or a canine hero – a virus to spread later to the cinema with *Lassie, Son of Lassie*, etc. and to television with *Daktari*. One such dog-drama – *Le Chien de Montargis* – led Goethe to resign his post as manager of the Court Theatre in Weimar rather than accept the opprobrium of having authorised its production. Chaos broke out in Dublin when the dog played truant for a fortnight! Another dog-drama of almost equal fame – or infamy depending on personal taste – was Douglas Jerrold's *Love Me, Love My Dog* with the vital canine role allotted to Hector. Such was its popularity that half the dog-owners in Britain named their own pet after him. Clearly, it is easy to dismiss this and other aspects of nineteenth century drama, not least the passion for archaeological exactness in stage settings that characterised the productions of Charles Kean, Sir Henry Irving, and David Belasco, as frivolous or pernicious. But, as I have already suggested, this is bound to be the case (with only rare exceptions) if we persist in using literary standards as our looking glass; for this is the wrong kind of mirror in the first place.

What the Victorian theatre truly reflects is the triumph of technology. To celebrate this adequately it needed startling crimes and necrophiliac horrors of Senecan dimensions, set in exotic landscapes and animated by soul-searing passions: for it was only such raw materials as these – whether in spectacular melodramas, Shakespeare, opera or ballet – that could enable scenic artists together with their machinists and lighting engineers to display their inventive skills to the full. Wagner's *Das Ring der Niebelungen* with its gold mine, its swimming Rhine-maidens, its dragon, its fire-encircled rock and the final collapse of Wotan's Palace epitomises this celebration of technology as applied to the theatre. Matching progress in optical science heralded the advent of motion pictures and with them a new era of dramatic art first in cinematic, and then in televisual form. It was only in the final decades of the Victorian era, however, that the theatre itself began to take note of these radical changes. Led by W.S. Gilbert, Robertson, Wilde and Shaw, it sought to purge itself of the turgid sentimentality and melodramatic sensationalism to which, in the course of its spectacular, technological advances, it had become addicted, and to inject restorative transfusions of witty debates on topics of genuine concern to society. Ironically, no sooner had it begun to set its own house in order in this way than theatre managements had to face up to the challenge of photographic reproduction of continuous action projected back to audiences on a screen. The prerogative, that up to this time had been unique to theatrical art, of communicating its messages to the brain *via* the ear and the eye simultaneously was thereby fragmented if not wholly shattered. For, as Shakespeare observed in *Coriolanus*, 'the eyes of the ignorant are more learned than their ears'; and accordingly, the mass, urban and suburban audience chose to follow the fortunes of the nascent movie industry where the spoken text was kept strictly subservient to the visual images, and so quit theatres just when their stages were seeking once again to accommodate serious talk and argument. This adjustment of British theatrical mirrors to reflect 'the very age and body of the time' thus coincided with the mass walk-out of the majority of its patrons. What the intellectual *avant-garde* that supported the Little Theatres and Theatre Clubs of the early twentieth century like the Stage Society and the Royal Court contributed in radical enthusiasm and aesthetic distinction, they wholly failed to compensate for numerically. What followed may thus be likened to a bath full of water with the plug pulled out and one tap left running. Box-office statistics and demolition contracts tell the rest of that story.

Many efforts have been made to reverse that flow, more especially since the end of the Second World War when even cinemas began to lose their patrons to television. Architects have been called in to re-define the relationship between the acting-space and the auditorium, and to translate or expand theatres into community centres for the enjoyment of the arts. Theatre history has been ransacked to find production styles that are conducive to stricter economy, greater intimacy and provision of a repertory with a recognisably educational bias. Clubs have been formed and Local Education Authorities enlisted to support them. Money has been deflected from rates and taxes to finance these changes and to improve artists' and technicians' salaries. The result is that, although still very much a minority activity, live-theatre in Britain is unquestionably more robust and faces the future with less evident anxiety than it has done for half a century.

When I cross-examine myself to find a reason for this surprising improvement in the health of that perennial hospital patient, 'the old sick man of the theatre', I notice that in its role of reflecting-glass it has been more successful in anatomising 'the very age and body of the time' in recent decades than has been the case for many generations. We have been more fortunate than most countries in the supply and dependability of our actors; exceptionally fortunate to be possessed of a larger and more talented succession of dramatists than any other country in the world; and fortunate to have seen so much public money injected into the saving, refurbishing and creation of playhouses. We have also learned from our medical advisers that if we wish to cure obesity we must accept dieting; and in theatrical terms this has meant that managers and producers have had to abandon scenic spectacle and large-cast plays. However, I do not myself believe that any of these factors, taken singly or collectively, can account for the revitalised condition in which our theatre finds itself today. The radical and dynamic change that is really responsible for this miracle is that almost total reversal of balance between metropolitan and provincial centres of gravity in theatrical life that has occurred during this century. And, as always when a shift of this order occurs in society, the root-cause of the resulting transformation is to be found in patterns of education. The slow, but steady, return of drama to a status of respectability in our academic institutions began in the last decade of the nineteenth century with the founding of the Oxford University Dramatic Society and the Amateur Dramatic Club in Cambridge. From these humble beginnings, nearly a century ago, drama has come to take its place alongside other disciplines in universities, in teacher training colleges and in schools: and from there it has been fed back through its students nationwide to the community at large in a variety of ways that were unimaginable even fifty years ago. As a result, we are slowly rediscovering that dramatic art depends for its vitality not on glittering chrome and neon palaces in capital cities, not on the glamour of stars and gala-nights, not on costly machinery and the armies of technicians needed to man them (for these are coming to be recognised as external manifestations of a sense of occasion, as idols, false gods, if worshipped for themselves) but on the simplest of all human relationships; a dialogue between an actor and his audience. This can take place anywhere; in a pub, a garden, a village hall, a street, a hospital ward, a classroom. Moreover, it is enjoyable if only because it destroys that most formidable of human enemies; the isolation of solitariness, of facing all life's problems alone. It is a relationship which combines giving with taking, and thus promotes sharing; a sharing of ideas; ideas that worry and disturb us; ideas that amuse us. This activity itself provokes enquiry, reflection, and a corresponding relaxing of tensions.

I am not going to indulge in the folly of assuming a prophet's mantle; but I am aware that, having posed the question at the outset of this lecture, 'Can our theatre survive this century, or is it bound by then to have been devoured by television?', it would be cowardly to side-step any sort of answer. Looking to the future then, I can at least give you my confident opinion that nothing can ever kill the theatre. For at amateur level, everyone is born possessed of a mimetic instinct: indeed, all our behaviour patterns are acquired by putting it to use from the cradle forward into adolescence. The recent trial in America of a youth charged with murder, whose defence council pleaded diminished responsibility attributable to constant exposure to television violence, is worth recalling in this context. In later life we may choose to use the mimetic instinct where and whenever two or three people are gathered together either privately as a mask behind which to hide our true feelings, or publicly as an instrument through which to explore aspects of the human condition that please us, that frighten us, or that we cannot fully understand. What our own theatre today proclaims, at times almost too stridently, is that civilised society as we have known it is disintegrating around us.

We may not like this, and complain that our playwrights have become too didactic and cynical: yet we behave like the Wicked Queen in *Snow White* if, seeing society beat a daily retreat into what, hitherto, was regarded as barbarism, and move into what seems like a new Dark Age of general illiteracy and blunted sensibilities, we then blame the mirror instead of 'the very age and body of the time' which so many 'sordid' or 'violent' modern plays simply reflect. Nevertheless, history insists that whenever playwrights have pressed their particular didactic enthusiasms to the point of becoming preachers, and in doing so have forgotten the festive nature of dramatic art, audiences have deserted them for forms of recreation better suited to their actual needs. This is the writing that television has been inscribing so insistently on our living room walls for the past twenty years and which our younger playwrights and actors, if they want a live theatre to work in for the rest of this century, will read, mark and inwardly digest. To be successful, a play is as dependent upon its form as on its content. If new and larger audiences are to be found to support the theatre, it is not ever larger subsidies wrenched from increasingly rebellious tax-payers that will secure them, but new stage-conventions and new rules of dramaturgy that ensure a genuine capacity to entertain as well as teach when endeavouring to reflect in a recognisable manner the order of things that comes to replace what we have known; just as medieval actors and playmakers found new ways of expressing and reflecting their own convictions centuries after the collapse of the ancient world and the theatre that had mirrored its beliefs; or just as the nineteenth century found new ways to reflect the changes wrought in the fabric of society by the industrial revolution.

To try to predict what these means will be would be pointless, since only the ways in which society itself evolves can determine that: it suffices, therefore, to possess enough faith in the resilience of human nature, in the boundless scope of human curiosity, and in the limitless drive of human will-power to prefer order to chaos, to feel assured of the ultimate survival of dramatic art and of its continued ability to build bridges of understanding between one human being and another, one community and another and last, but far from least, between past civilisations and their successors.

4th November, 1977

LECTURE VII

Experiment and Experience in the Arts

Ernst Gombrich

It seems to me a pleasant fancy to imagine that no event in this world ever disappears without trace, and that even the words spoken in a particular room continue to reverberate, ever more slightly, long after their audible echoes have faded. If that were true a supersensitive instrument might still be able to pick up the resonance of words spoken in this very hall a little less than 142 years ago in what I imagine to have been a vigorous Suffolk accent, very different from mine.

'Painting is a science' – you would hear the voice say – 'and should be pursued as an inquiry into the laws of Nature. Why, then, may not landscape painting be considered as a branch of natural philosophy, of which pictures are but the experiments'? The artist who was thus appealing to the *genius loci* of this place was John Constable (Fig 11) and the occasion the last of four lectures he gave at the Royal Institution in April 1836, for which the invitation is still preserved in its library.[1]

I regard it as peculiarly fitting, therefore, that it was this Institution which has been selected by the Richard Bradford Trust for a series of Seven Lectures on the Influence of the Arts and of Scientific Thought on Human Progress, of which this is the last. One of their aims should be, as I read in the accompanying memorandum, 'to test the theory that there is a close analogy between the processes involved in the development of a scientific hypothesis and of an artistic creation', the analogy in fact, to which Constable appealed in the words I have just quoted. It is for this reason that I should like as far as this is possible within this compass, to take the notion of the experiment as my guide in looking at the relation between the sciences and the arts. The role which experiment plays in science has been outlined by both speakers on scientific topics in this series. To quote Dr David Samuel's lecture on brain research: 'Scientists must remember that though they should be audacious in seeking hypotheses, they must be meticulous in experimentation'. The reason for this injunction becomes clear in Sir Peter Medawar's exposition of the Philosophy of Karl Popper where he states succinctly that 'Acts undertaken to test a hypothesis are referred to as experiments'. Such tests, which should be public and repeatable, determine the scientist whether to hold on to his 'audacious' hypothesis, or modify, if not abandon it. In either case we may speak of the success of the experiment, for the refutation of a wrong hypothesis is vital for the progress of science, leaving the path open for further searching and probing. In the language of engineering this approximation to a goal through the elimination of errors is known as 'negative feedback'. Can we observe a similar mechanism in the evolution of the arts?

It is suggested in the memorandum that a work of art is actually 'subject to verification or falsification very much like a scientific theory,' the test being 'whether the work is accepted or rejected by the large majority of those to whom it is submitted'. I shall have to come to further amplifications and qualifications of this principle later. For the moment it must be granted that it is certainly possible to see the evolution of art also in the light of 'negative feedback'. But put in this form the answer may be insufficiently specific to be tested in its turn. For we have learned from Darwin that all evolution can be seen in terms of the survival of the fittest and a good case can always be made out for applying this approach also to the history of civilisation.

Whether you think of technology, or religion or of social institutions, you have at any moment of time a possible spread of variants of which most are still-born while others are taken up and survive. Technology seems the most obvious example. It is tempting and easy by hindsight to say that bronze tools were bound to prove better than those of stone, and iron ones better still, though there probably were moments when it was not clear yet which technology had more advantages on its side. It is harder to say why one religion or sect overtook another or why certain institutions were so eagerly adopted by one culture and rejected by others. Yet I would agree that artistic developments can also be viewed from that angle and that we can say by hindsight that Egyptian art was just the right art for Egyptian religion and society, and the same must apply to Sumerian, Minoan or Mexican art, the examples chosen by Jacquetta Hawkes in her lecture in this series. Whatever we see as the ultimate causes of the formation of these contrasting styles – and I am not sure that we can rest satisfied with the answers which have been proposed – we may still agree that the images adopted by a particular culture must have evolved from something like selective pressure. Certain qualities were found more appealing or intuitively more in accord with the desired aims than others and these were codified as conventions. What was said in that lecture about Sumerian art also applies to other ancient civilisations. 'All works were created for cultic purposes to celebrate the divinities, or the Kings who were their stewards on earth'. Maybe 'to placate' would be clearer than 'to celebrate', for ultimately it was neither their beauty nor their impressiveness which mattered but their efficacy in securing the favours of the higher powers, a good harvest, victory in war, the discomfiture of the enemy, and perhaps mercy for the departed souls. It is in the nature of such a ritualistic conception that the very idea of wanting to test the power of these artistic enterprises would be blasphemous. In rites, chants or the construction of amulets and spells you follow precedent in the confidence that they will work their magic. Hence an attitude like the Egyptian was, in the words of the lecture, 'almost bound to lead to highly traditional arts, with a fixed symbolic meaning in every attitude, action and object'. Indeed the arts which are 'completely enmeshed in their religious mythologies' could never be tested from the inside as it were. If you seek for experiments you would have to go to the followers of other religions, to the prophets, saints or missionaries who wanted to convince the populace that their idols are merely sticks and stones by smashing them with impunity. Jacquetta Hawkes has told us of the astounding episode in Egypt in which Akhenaten caused the names of all gods save the sundisk to be erased and instructed the sculptor to depart from the ritualistic style of precedent. The difficulty of interpreting a stylistic revolution over such a distance of time makes one wonder whether the realism he demanded in the portrayal of his person and of his family was a symptom of humility or of pride – if, indeed, these terms can apply to such an exceptional situation. To what extent this style would have consolidated into a set of conventions if Aten's religion had endured we can never know. At any rate, not all students of the period regard it as an analogy to our Romantic movement in art.[2] If, as she assumes, the art was profoundly shocking to conservatives in Egypt it was, no doubt, because they feared dire consequences from the non-observance of sanctified rules.

The replacement of this attitude by critical awareness is one of the crucial elements in what Jacquetta Hawkes justly describes as that greatest revolution in human thinking achieved by the Greeks which, in her words, 'not only initiated science but also vastly increased the conscious, thoughtful aspect of art'. Professor Kitto in his lecture on the Greeks rightly warns us to watch our terms here, for it is always risky to assume that the Greeks operated with the same concepts of religion, science and art as we do. They had no word for art in our sense, because *techne* meant skill in any aspect of culture, in fortification no less than in image making. But maybe it is precisely because all arts were seen as skills that the decisive element entered the situation which explains much of that increased consciousness I have mentioned – I mean criticism. To criticise is to make distinctions and the critic is the professional fault finder. Remembering what we have heard about the importance of negative feedback this detection of mistakes is no minor function. In any case the need for the critic who could articulate his judgement arose in many aspects of Greek life, for the Greeks – as Jakob Burckhardt pointed out – had a passion for public competitions.

It is no mere pun to remind you here how close the term contest is related to the term test; there were tests and contests everywhere not only at the Olympic games but at any village feast. Now the idea of skill is only applicable to art if you look at art from an instrumental point of view as trying to achieve a particular end, and this was certainly the Greek approach.

I mentioned that in ancient cultures the implicit end of art was efficacy, the licensed magic of religious rites. Speaking schematically one might say that this element of magic is still very much in evidence in the Greek view of the arts. There is indeed an easy transition from the idea of a religious incantation to that of the spell cast by art. A thinker such as Plato judged art, dancing, poetry, music and to some extent image making by its effects. If I may be allowed another oversimplification I would describe this development as a transition from religion to medicine. Art for Plato is like a drug, its effects can be rousing or tranquillizing, invigorating or effeminating and it is precisely for this reason that in his view it had to be strictly controlled by the state. It is well known that he looked back with nostalgia and approbation to the ancient Egyptians who had, he thought, exercised such a control for thousands of years and had prevented innovation as dangerous and subversive. You must not tamper with such powerful agents. Homer should be banished from the ideal Republic, politely, it is true, but banished all the same. Plato is the predecessor of those enemies of the stage mentioned by Professor Wickham in his lecture and those of us who fear the corrupting influence of violence on TV, a possibility conceded by Lord Clark.

But clearly there is a vital difference between censorship and criticism. Greek culture opted for the second rather than the first. The distinctions of the critic cannot be based on blanket approval or disapproval but on the verdict of good, better and best. We must not forget in this context that Greek drama, the central theme not only of Professor Kitto's lecture but also of the Bradford statement, evolved in and through competitions. These solemn re-enactments of communal myths performed at the festival of Dionysos differed from other communal rites precisely because they did not encourage but rejected artistic conservativism. We have a schematic picture of the resulting development in Aristotle's Poetics where he tells us that Thespis has only one actor, Aeschylus two, Sophocles three, and so on. But these developing means were harnessed in his view to a desired effect, which he defines in half religious, half medical terms as *katharsis*, or purgation of the passion. Of course this is not only a matter of technical means, but of the most compelling interpretation of the myth. We have four versions of the Electra story[3] and each can be seen as an attempt to improve on the earlier. It is not far fetched to speak here of experimentation, though not yet of experiment. By experimentation I mean the attempt to look for alternatives which are more effective, more suitable for the achievement of the desired end.

Whatever may be true of the playwright, there is no doubt that this spirit of experimentation pervaded the art of image making during the Greek revolution. It looks as if the painters and sculptors were set to solve a predetermined problem, the problem which is traditionally described as *mimesis*, the correct representation of Nature, but which I

11 Daniel Maclise, Portrait of Constable painting, *c.* 1831

12 and 13
Klytemnestra held back (above) and Orestes killing Aegisthus, from a red figure pelike by the Berlin painter

4 The Purification of Orestes: red figure bell-krater from Lucania, by the Eumenides painter

15 Orestes and Pylades in front of King Thoas, mural from Pompeii

16 Detail of the 'Alexander Mosaic' from Pompeii

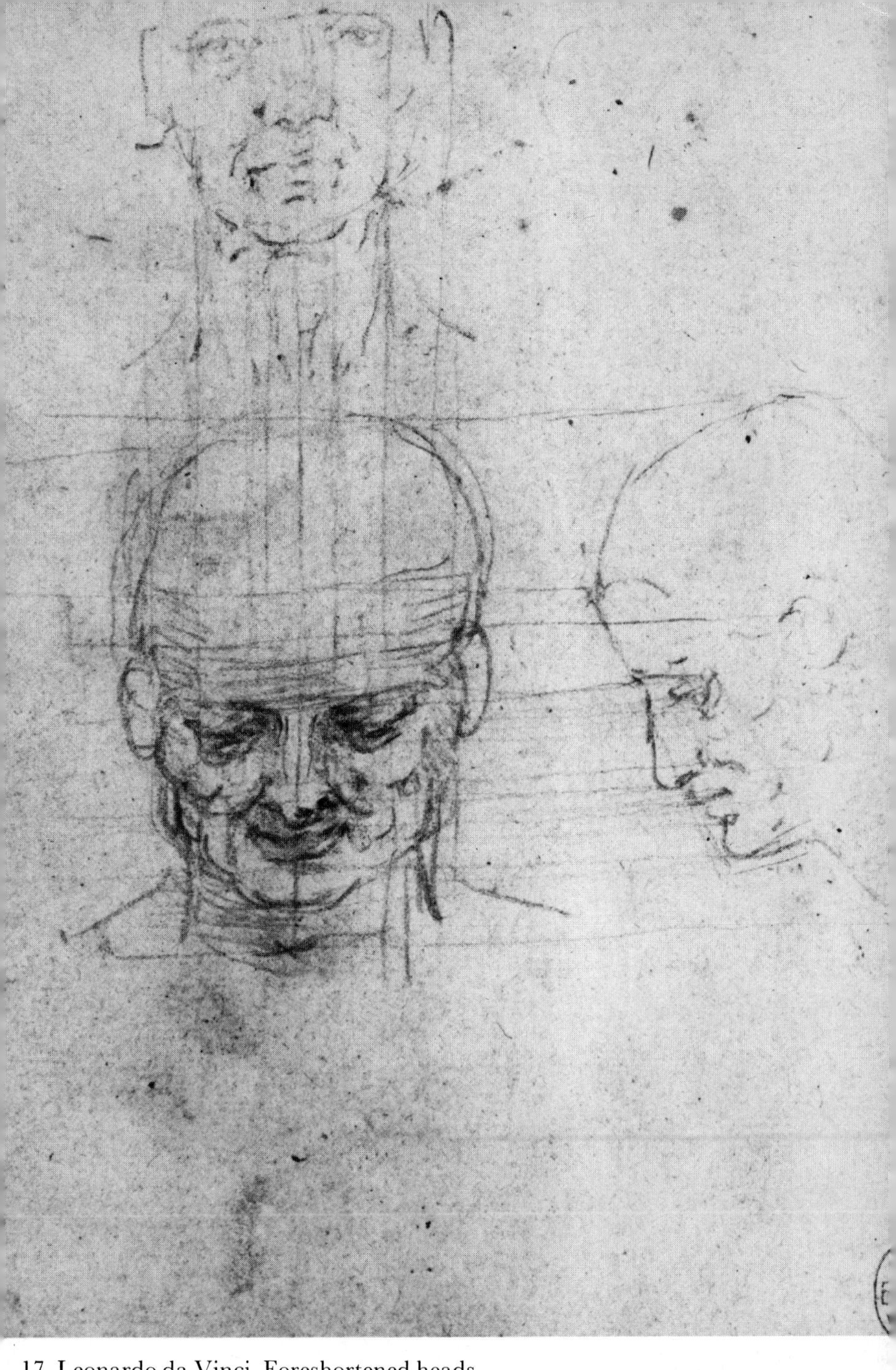

17 Leonardo da Vinci, Foreshortened heads

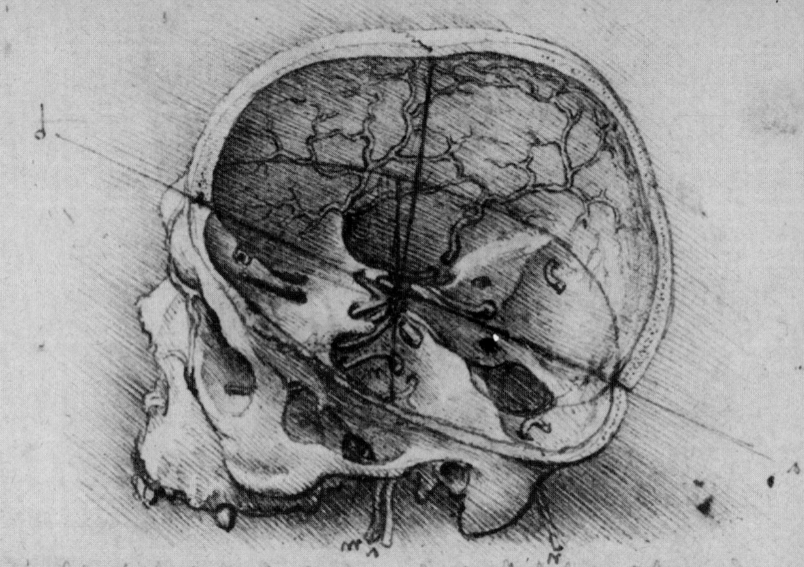

18 Leonardo da Vinci, Section of a skull

19 Francesco Marmitta, Madonna and Child, drawing

20 Leonardo da Vinci, Madonna and Child, drawing

21 Leonardo da Vinci, Studies

22 Leonardo da Vinci, The Madonna of the Rocks

prefer to connect with the wish to turn the beholder into an imaginary eyewitness of the mythical events.[4] It was for this reason that Greek masters since the late sixth century attempted with ever more intensity to struggle free from the conventions of art which had, for instance, governed the Egyptians. A redfigured vase in Vienna (Figs 12 and 13) of around 500 B.C. shows us a scene from the story of Orestes which Sophocles does not show on the stage, the killing of Aegisthus by Orestes, all marked by names. It is Klytemnestra who rushes to the aid of her dying husband with an axe, but restrained by one Thalbytios, while Chrysothemis rather than Electra assists the terrible scene. What may strike us first is perhaps the archaic rigidity of the style, the way the bodies are still twisted somewhat like Egyptian figures to show the trunk from in front and the head invariably in profile, but this impression is belied by the rendering of the hands and feet. The left hand of Aegisthus grips the foot of the throne in vain while his right attempts desperately to restrain the arm of Orestes.

We are fortunate in being able to document the spirit of competition which inspired the experimentation of these vase painters. There is a vase in Munich by the painter Euthymides which bears the famous inscription: *Hos oudepote Euphronios*, Euphronios (a rival vase painter) 'has nowhere done this'. It is clear from the picture to which this proud boast is attached that 'this' must mean the mastery of complex movements seen in foreshortening. But even without this written evidence we have the proof in the further development of Greek art that professional fault-finding led to progressive solutions. Two generations later another episode from the Oresteia – the purification at Delphi (Fig. 14) – is represented with the freedom that was the result of further experimentation. Some heads are in profile, some *en face*, and the impression of the relaxed ease of movement that characterises the whole scene is due to the mastery of the skill of foreshortening. A painting in Pompeii, reflecting a Greek original of the fourth century, Fig

15, shows another scene. Orestes and Pylades in Tauris in front of the King, about to be sacrificed on the altar before they are recognised by their long lost sister Iphigenia, the priestess. One does not have to enlarge on the fall of light and the convincing rendering of space, but this only provides the setting for the expressiveness of the two beautiful nudes. Even those who prefer the vigour of the earlier style must acknowledge that what might be called 'directed experimentation' had led Greek artists towards the solution of its problem of how to evoke for us a convincing and moving vision of this episode.

Now it so happens that the effectiveness of these means of art was also reported to have been tested by experiments – in fact by animal experiments. The story concerns the mastery of representing life-like horses of which the Alexander mosaic in Naples may convey some idea (Fig 16). Not that we need believe the anecdote told by Pliny in his *Natural History* (XXXV, 95) but it is symptomatic all the same.

We read that the greatest painter of that period, the famous Apelles, competed with a rival as to who could do a better painting of a horse, and when he suspected his rivals of some kind of foul play in securing the prize, he asked for horses to be taken to the exhibition. Lo and behold, the horses only whinnied in front of his own painting while those of rivals left them cold. 'Afterwards,' remarks our source, drily and unconvincingly, 'this experimental test of art, (*experimentum artis*) was always used.' In other words the tests of public repeatability were applied – or so they say.

But I don't want to give the impression that art criticism in the ancient world was left exclusively to horses. There is a story in Lucian about the famous precursor of Apelles, the painter Zeuxis, which brings in a very different element. Zeuxis had painted a picture of a family of Centaurs. The painting made an immediate hit because of the charm and the originality of the subject matter. But the very compliments he received on that score annoyed the artist. 'Pack it up and take it home' he says in the dialogue *Zeuxis and*

Antiochus, 'the people are delighted with the earthy part of my work, novelty of subject goes for more with them than truth of rendering. Artistic merit is of no account'.

Note that the contrast between the judgement of the crowd and the approval of the discerning, between the horse and the connoisseur, worried the ancients as it still worries us. It is a problem that does not really bother the scientist for it is obvious that to evaluate a scientific experiment you have to understand its aim. In art this kind of distinction is now labelled as élitism, an ugly vogue word which still points to a real problem as long as you look at art as an instrument designed to create a psychological effect.

In this respect the model art for the ancient world was not painting nor even poetry but oratory, the art of gaining friends and influencing people which was of such vital importance in the ancient democracies. Accordingly the art of rhetoric in all its aspects was the subject of the most sophisticated analysis in ancient literature. Cicero, the leading Roman practitioner of this art has left us his reflections on these matters in many of his writings. In one of them, the dialogue *Brutus*, he traces the rise of oratorical skills in the Roman Republic and carefully compares the means used by the spellbinders of yore. And yet, he admits, you don't really have to know all these technicalities to make up your mind about the quality of a speaker in a court of law. In fact you don't even have to listen to the speaker. A casual glance will tell the listener whether the orator who is on his feet knows his job: If one of the judges is seen to yawn or to talk to his neighbour there is no orator present who can, in Cicero's words, 'play on the minds of the court, as the hand of the musician plays on the strings' (*Brutus*, 200). What the horse was to the picture of Apelles the jury is to the orator. It is the effect alone which counts as a test, and audience ratings are really all that matters in the end. But Cicero warns us also not to generalise too far. There is a court of appeal against the verdict of the crowd. He tells of a poet who read an epic which he composed to an assembled audience, only to find that all listeners had left, except for Plato. 'I shall go on reading all the same,' he said, 'for me the one Plato counts one hundred thousand' (*Brutus*, 191). Cicero believed in what the media call minority programmes. More than that, he implied in an even more moving anecdote that even Plato is not the ultimate court of appeal. When the playing of a flautist was coldly received by an audience his teacher pleaded: 'sing for me and for the muses' (*Mihi cane, et Musis*) (*Brutus*, 187). The Muses remain in heaven even when the gods have departed. Those of us who believe in the Muses as the final arbiters set less store by horses, by judges, and even by Plato. But I am the first to admit that such a faith creates its own problem precisely because we have no direct line to the Muses and cannot test their reaction. No wonder that the objective canons of science promised for a time to offer a valid substitute, at least in the visual arts.

In attempting to explain this appeal to science, for which Constable served me as my first witness, I must point to the remarkable similarities between the scenario I have described and the one that followed. There was a tendency in the Christian art of the West, and even more so of the East, to revert to a ritualistic concept of image making, but the purpose assigned to the visual image in the West potentially favoured a return to dramatic evocation. Briefly, pictures could be used to impress on the minds of the faithful the teachings of the Church. Professor Wickham's lecture dealt in this context with the role of the miracle plays and here, as in ancient Greece, the stage and the arts followed a parallel course. In the later Middle Ages, sculpture and painting began to rival the popular preachers in their efforts to evoke the stories of the scriptures or the miracles of the Saints as vividly and movingly as possible. We are used to connecting this new spirit of exploration and experimentation with the idea of the Renaissance, the return to classical standards, but it would be a mistake to omit from this epic of conquest the contribution made by the Gothic North.

What was distinctive of the Italian Renaissance was the

way the aid of science was invoked to create a semblance of reality. Even the most moving picture could be found to jar if attention fastened on inconsistencies in spatial relations, indeed the more so the closer they otherwise came to a convincing setting. To correct these inconsistencies an understanding was needed of the laws of optics which govern the process of vision, in other words of perspective, and it is here that we learn of the first experimental demonstration connected with painting. The demonstration was due to the great Florentine architect Filippo Brunelleschi who solved the problem on the basis of Euclidian geometry. His experiment took the form of a peepshow – alas no longer extant – in which you could see through the eyehole a view of the Florentine Baptistry as seen through the open doors of the Cathedral. It showed that it was possible to work out theoretically what aspect of any object could be seen from any particular point in space.[5]

But a knowledge of the laws of projective geometry cannot help the painter in creating that desired semblance of reality unless he also has a knowledge of the structure of the objects he wishes to include in his picture – most of all, of course, the structure of the human body. Anatomy, therefore, became the other branch of science eagerly studied by the artists of the Italian Renaissance. It was in fifteenth century Florence, then, that that tradition was born to which John Constable appealed in his description of painting as a science which should be pursued as an enquiry into the Laws of Nature, and it was that greatest of Florentines, Leonardo da Vinci, in whose writings this conviction is given the most vigorous expression. There is no doubt that Leonardo's investigations into so many aspects of Nature took their starting point from the demand for objective criteria in the rendering of structure and appearance (Figs 17 and 18). Too often, he knew, we deceive ourselves if we rely on our impressions rather than on measurement. Thus he notes for the instruction of budding painters who should not trust traditional routine:

'That painting is the most praiseworthy that conforms most skilfully to the object it represents, and I say this to confound the painters who want to improve the work of Nature, for instance those who represent a one year old boy with the proportions of a man of thirty, giving the body eight headlength instead of five (Fig 19). They have so often committed this error and seen it committed so often, that they have become used to it, and the habit has lodged itself so deeply in their corrupt judgement, that they make themselves believe that Nature, or whoever imitates Nature, are gravely in the wrong when they deviate from this practice.'[6]

One senses that Leonardo who liked to represent his Christchild with a very large babyhead (Fig 20) encountered the criticism of traditional painters. We shall find that the need for measurement and observation becomes henceforward linked with a warning against the lazy habits and prejudices of artistic bumbledom.

For Leonardo, of course, the need to test, to observe, to experiment, not to take anything for granted, had become second nature. It is not always possible in his writings to tell when he uses the word '*sperienza*' whether he is talking of 'experience' or of experiment, but there are enough passages which show unambiguously that he had a clear notion of the modern meaning of the term experiment. Thus there are two pages of the Codex Forster, in the Victoria and Albert Museum (pp II 2f, 135r and 67v) in which he studies the laws of suspension and leverage, one arrangement being marked '*sperimentato*', (tested), the other '*nonn isperimentato*', (not tested).

Here we have a real experiment in the scientific sense, but in studying Leonardo's writings we soon realise that he extended the principle of experimentation far beyond that of objective testing. Confronted with any problem he liked to resort to a systematic permutation of the elements as if he wanted first to make sure that he had not overlooked a single possibility.[7] We find him applying this method of experimentation rather than experiment in his geometrical and architectural studies no less than in his compositional sketches. These are not simply preparatory stages for the finished work, they are 'studies' in the modern sense of the term, attempts to seek the best solutions for an effective gesture or a balanced grouping. What is at stake, in other words, is no longer the achievement of fidelity to Nature; rather it is the realisation of the artist's own vision which is pursued through trial and error. We are so used to this procedure that we may fail to see the psychological puzzle it presents. Why should an artist put something on paper only to reject it?

The explanation was furnished by Giorgio Vasari in the middle of the sixteenth century. Vasari was a Tuscan, like Leonardo, and had gone to Venice for a time, where he must have engaged in many debates about art. At any rate he alleged, not without gross exaggeration, that Venetian painters did not believe in the value of making studies on paper. None other than Titian is reported to have believed that to paint immediately in colour was the best method. 'He failed to realise' Vasari claims, in his life of Titian, 'that it is necessary to anyone who wants to arrive at a good composition and to adjust his inventions, first to draw them in different ways on paper so as to see how it all goes together. The reason is that the mind can neither perceive nor perfectly imagine such inventions within itself unless it opens up and shows its conceptions to the corporal eyes which aid it to arrive at a good judgement.'

Here we have the idea of experimentation, of negative feedback applied to the artist himself. He is his own experimental guineapig submitting the inventions of his mind to the critical judgement of his eyes. The more an artist ventures into the unknown by abandoning the well tried methods of tradition the more vital is this procedure likely to be. Indeed it may be claimed that this discipline of self-criticism has become the most precious heritage of Western art.

It is all the more important, in my view, to clarify in what respect it differs from the principle of scientific experimentation. It differs, I would claim, because the goal which the artist seeks with such self-critical persistence is not a true proposition (as in science), but a psychological effect. Such effects can be discussed, but they cannot be demonstrated.

We need not stray further than the art of Leonardo to give substance to these general considerations. For however intent the master may have been in making his pictures conform to visual reality, there are limits to this aspiration. Let us grant (to avoid misunderstanding) that it is possible to make a fascimile of any object in three dimensions or of any coloured surface, but who would dare to say whether the *Virgin of the Rocks* (Fig 22) matches objectively what we would see when entering such a cave? It would depend on our adaptation to the darkness and many other factors, but in normal conditions there never could be an objective correspondence, for there is a limit to which pigments can match light.

The effect which the painter here seeks, therefore, is an impression of equivalence. He has to grope his way to come as close as possible, but he has only his own 'corporeal eyes' to tell him how far he has realised his vision. Other eyes may or may not be satisfied. Once more Vasari, who is so frequently maligned, shows a remarkable understanding of the situation when he comments on the dark appearance of Leonardo's paintings. 'He went ever more deeply into the shadows and searched for blacks yet darker than other blacks in order to achieve brightness and luminosity, till his paintings look like renderings of night scenes rather than daylight: yet he did it all to achieve more relief and to attain perfection in art' (Fig 23).

Here we have a good example of an aesthetic problem, the solution of which is hard to measure and quantify. All we can say is that, like Vasari, other Renaissance painters were not convinced that the method really worked, while in the seventeenth century further experimentation led from Caravaggio to Rembrandt.

It fits my books that Constable came to speak of this very problem of matching light by paint when in his second discourse he discussed the works of the 17th Century master Claude Lorrain (Fig 24), who, as he said, 'carried landscape to perfection, that is *human perfection*'. 'No doubt' – he continued – 'the greatest masters considered their best efforts but as experiments, and perhaps as experiments that had failed when compared with their hopes, their wishes, and with what they saw in Nature. When we speak of the perfection of art, we must recollect what the materials are with which a painter contends with Nature. For the light of the sun he has but patent yellow and white lead, for the darkest shade, umber or soot'.[8] In these experiments, as Constable rightly stressed, 'Claude, though one of the most isolated of all painters, was still legitimately connected with the chain of art'. He could not have existed without his predecessors. 'He was, therefore, not a *self-taught artist*, not did there ever exist a great artist who was so. A *self-taught artist* is one taught by a very ignorant person'.

What Constable singled out in Claude was his integrity, 'there is no evasion', as he puts it, and only in his declining years were 'his former habits of incessant observation of Nature . . . departing from him'. For with all his emphasis on tradition Constable's main critical target was what he called 'mannerism', 'the imitation of preceding styles with little reference to nature'. He singled out the Italianate Dutch masters such as Both and Berghem (Fig 25) for strictures, in whose works 'all the commonplace rules of art are observed; their manipulation is dextrous and their finish plausible' but their truth is second hand and their reputation merely kept up by dealers. After the lecture – we hear from Leslie – a member of the public said to Constable: 'I suppose I had better sell my Berghems', to which he replied 'No sir, that will only continue the mischief, *burn them*'.

It is not hard to see what the conception of science was which influenced Constable in his lectures at the Royal

Institution. Even if he had been less fond of quoting Bacon we could tell that his view of human progress was Baconian. The only enemy of true knowledge is prejudice, laziness. As Popper has suggested in an important lecture,[9] in Bacon's view it is our sinful nature which prevents us from seeing the truth God has placed in front of our eyes. Like the scientist the landscape painter must not take received opinion as truth. He must apply himself as an investigator of natural phenomena and ceaselessly observe and record them. It is likely that when Constable suggested that pictures are experiments he was thinking of such observational records in the service of 'induction' rather than of the testing of theories, though the two were less sharply distinguished then than they have become since. His art confirms this interpretation. He noted with marvellous precision the fleeting effects of light and of weather in the landscapes he painted (Fig 26). We know that he passed on this conception of the painter's mission to his friend and biographer C.R. Leslie in whose *Handbook for young Painters* we find it even more explicitly stated. Leslie mentions there (pp. 259–260) celestial phenomena 'not yet made tributary to Art; the lunar rainbow, for instance, and the aurora borealis. . . . There is also a beautiful appearance in calm weather, when large masses of bright clouds are reflected in broad columns of light on the sea, just as the sun throws his pillar of fire below him. I may be mistaken, but I cannot recollect this in a picture, constant as its appearance is in Nature. . . . The truth is, we go on painting the things that others and ourselves have painted before. . . . Now and then an original painter adds something new and beautiful, but the most original might be more so, were it not for that natural indolence that makes even such too easily content to rest in what has been done'.

Natural indolence, sloth; the most eloquent witness to this faith in art as an ally of science in search for the truth is not Leslie nor even Constable but John Ruskin. One would love to know whether it is at all possible that Ruskin, then a young man of seventeen about to go to Oxford was among the two hundred people who came to Constable's lectures at the Royal Institution. He had taken drawing lessons from Harding and had actually written or was about to write an essay in defence of Turner.

It was this theme which developed seven years later into *Modern Painters*, the most ambitious work of scientific art criticism ever attempted. The main sections of the book consist of a survey of natural phenomena, 'Of Truth of Skies', 'Of Truth of Clouds', of mountains, of water and vegetation described and analysed in marvellous detail, always with the purpose of demonstrating that even the most famous landscape painters of the past had fallen short of perceiving the truth and that they had all been surpassed by Turner (Fig 28). In the first chapter of his section 'Of Truth of Water' he lets himself go in his contempt for such famous marine painters as Backhuysen (Fig 29) and Vandevelde (Fig 27) – who 'are thought to have painted the sea, and the uninterpreted streams and maligned sea hiss shame upon us from all their rocky beds and hollow shores'. How can he convince his readers that the water painting of the elder masters is so execrable? 'When I find they can even endure the sight of a Backhuysen on their room walls, (I speak seriously) it makes me hopeless at once. . . . Another discouraging point is', he goes on, 'that I cannot catch a wave, nor Daguerreotype it, and so there is no coming to pure demonstration'. He was writing in 1843, and there was no way yet of satisfying his search for an objective record of appearances by any mechanical means. It was painting or nothing.

Even eleven years later it was possible to catch a wave on the photographic plate via a new process (Fig 30), and, if so desired, to compare it with Turner. The new way in which science had taken a hand in the process of image making must have posed a serious problem to the outlook we have considered. I have always found it significant and moving that Constable's son became concerned with photography.

When Ruskin came to discuss the relation of science and art in 1871 he was clearly on the defensive. It appears that Thomas Huxley had drily remarked to him that recording Nature was a purely mechanical matter, and Ruskin felt compelled to explain why it was not. It was not the phenomena themselves the artist analysed, as the scientist did, but their appearance, their effects on him. In doing so scientific knowledge might in fact hinder rather than help him, for it created intellectual prejudices which prevented him from recording what he really saw rather than what he expected to see.

Like Constable he appeals to the notion of experiment but his meaning is less apparent. 'Experiments in art' – he writes – 'are difficult and often take years to try'.[10] Perhaps he was thinking also of his friends the Pre-Raphaelites, equally bent on combatting the corruptions of routine. Indeed one of them, William Holman Hunt, in discussing the place in the movement of the utmost elaboration in painting puts it on record that 'I have retained later than either of my companions did, the restrained handling of an experimentalist'.[11]

But the experiments in art which attracted most attention were not made in Ruskin's England, but in France with that revolution which was marked by the clash about Manet's *Déjeuner sur l'herbe* shown and derided in the Salon des Refusés of 1863, and which achieved its fruition in the triumph of Impressionism.

The epic of this struggle forms the centre of Zola's novel *l'Oeuvre*, published in 1885 but containing recollections of earlier episodes the author must have witnessed. The novel has a bad name among lovers of art for Zola incorporated traces of his friend Cézanne in the figure of his tragic hero who commits suicide in the end.[12] But whatever we may think of Zola's act of disloyalty, the novel remains priceless as a document because it illustrates the enormous impact which the rise and prestige of science had achieved at that time on writers and artists alike. *L'Oeuvre* is in fact part of that enormous series of novels Zola wrote to demonstrate the workings of heredity in one family, the Rougon-Macquarts; the series was to exemplify his programme of experimental novel, the *Roman Expérimental* he had outlined in 1880. He had borrowed this challenging designation from one of the most famous landmarks in the history of medicine, Claude Bernard's *Introduction à la médicine expérimentale*, by the great physiologist who had discovered the vaso-motoric nerves, the function of the pancreas and that of the liver. But could a novel rival these achievements? It could, in Zola's view, if it accepted the findings of science about the forces determining human life and behaviour. Placing two characters together in a given situation could be compared to putting two chemicals into a proving glass. The result was predetermined by known causes. It is interesting to recall that a chemical metaphor was used in Constable's time by Goethe in his novel *Elective Affinities*. An even harsher materialistic determinism governs the plot of Büchner's unfinished play *Wozzeck* which centres on a dietary experiment performed by the doctor on the helpless downtrodden hero till he commits murder. If Zola's hero Claude Lantier suffers a similar fate it was because the author had also placed him in a hopeless plight. It was the plight of the painters around 1865 as the author saw it from the perspective of 1885. 'Science is today the only possible source,' – says Claude – 'but what should we take from it, how can we march with it?'[13] Yet in his buoyant mood he has no doubt of his mission and its value. Declaiming against the academic routine of copying the masterpieces in the Louvre, he vows he would rather cut off his thumb than return there and spoil his eyes in this way which for ever blinds one to the sight of the world we live in. Was not a bunch of carrots, yes, a bunch of carrots studied directly and painted naively with the personal note in which you saw it worth all the interminable confections of the schools? The time would come when a single original carrot would be pregnant with revolution. – Why revolution? Because, we understand by now, it would reveal a truth we usually hide from ourselves and from others. It would tear away the veil of sloth and prejudice and would be incorruptible like science. Of course even on the most charitable interpretation of this wild talk it is not the painted carrot which would do that. We have had and can have any number of scientific images of carrots by now, photographs, holograms or scannings by the electron-microscope; they have not really brought us much nearer to utopia, but then this is not what Zola's hero means. What he considers pregnant with revolution is the simple act of daring to use one's eyes. The act of painting an unconventional motif in an unconventional way, as Manet did in his still lifes, (Fig 31) becomes a metaphor, a symbol of a social and political attitude, of non-conformism and defiance of conventions – but it is the authority of science which gives the artist courage to perform this act.

In the eighties when Zola wrote this novel the search for this authority has gone further. I am alluding to the birth of what is known as Post-Impressionism in the art of Georges Seurat. What was believed to be his programme can be gathered from the letter which his friend and admirer the great Camille Pissarro wrote to his dealer Durand Ruel in 1886. 'To look for the modern synthesis by the means based on science, which are grounded on the theory of colours discovered by Chevreul, and later on the experiments of Maxwell and the measurements of O.N. Rood.'[14]

Of those mentioned, Chevreul was the acknowledged authority on the perception of colour, the discoverer of what is known as simultaneous contrast and of similar effects of colour interaction. The famous Maxwell, who became the first Professor of Experimental Physics in Cambridge is here quoted for his studies of colour vision including colour blindness, and the American O.N. Rood was the author of a book on *The Scientific Theory of Colour*, which had recently been translated into French. The artists were thoroughly up to date and they hoped to use these new theories for the solution of the problem of which I have spoken in relation to Leonardo and Claude – the problem of matching the effects of light by means of pigments. If you mix pigments they begin to look muddy and so it seemed more promising to build up a mosaic of dots in primary colours on the canvas and leave it to the eye to combine their effects without loss of luminosity. Seurat was a great artist and the very effort of analysis which compelled him to deny local colour and even outlines made him search for compensatory moves of simplification which proved of great interest. Even so, the hoped for effect did not quite materialise. In fact Professor Weale has recently told us that from the point of view of optics the experiment was doomed because it was based on an oversimplification.[15] An investigation of the interdependence of colour and size published in 1894, three years after Seurat's premature death, could have explained to him why the luminosity he hoped for disappeared at a distance from the canvas when the apparent size of the dots dwindled. Pissarro himself was soon disillusioned and thought he had wasted his time. Not that there was no profit to be gained from the scientific study of colour interaction. One of Seurat's friends was working towards a method of colour photography which, of course, draws on these principles, but then we must not forget that the coloured snapshots we show from our holidays borrow their luminosity from the strong lamp in the lantern, as does colour television.

There has been much argument about the effect of the photography explosion on the development of painting in this century. I personally have little doubt that it was crucial. For in a sense painting had lost what biologists call its ecological niche. Having been threatened before by the decline of religious art, it now had to look for an alternative function where science could not compete. The *volte face* was facilitated by a change in the climate of opinion towards the end of the nineteenth century. Around 1890 the French critic Aurier wrote that many scientists and scholars were discouraged. 'They realise' – he writes – 'that this experimental science of which they were so proud is a thousand times less certain than the most bizarre theogony or the maddest metaphysical reverie'.[16] Having been told in this series that it is not certainty which science claims, but a method to detect errors, we may be unimpressed, but we have to note the paradox that this method minus its function was appropriated as never before by artists, musicians and writers in search of alternatives. The word experiment became a vogue word to be used indiscriminately for any departure from tradition, any unconventional enterprise on the stage, in dancing, in poetry or in the application of new media.[16] So thoroughly has the term entered common parlance that I was amused to read in the notice of a recent novel in *The Times* (15.10.1977) that it suffered from a 'reach-me-down experimentalism'. The fact is that artists and critics could not but remain impressed by the triumphal progress of science in the twentieth as in the nineteenth century, and that all the more as they too continued to embrace the philosophy of progress. There are few twentieth century movements whose champions have not appealed to the example of contemporary science. Kandinsky wanted to link his abstract painting with the splitting of the atom, which to him symbolised the disappearance of solid matter.[18] Cubism has been coupled again and again with Einstein's theory of relativity in statements which it would be uncharitable to quote. Surrealism has made play with the Freudian unconscious, though Freud remained unimpressed; the game goes on with structuralism, linguistics or what have you, but needless to say these are efforts to profit from the prestige of certain scientific fashions rather than experiments testing their validity.

There are exceptions such as the exploitation of flicker effects in what is called 'op' art by artists such as Bridget Riley and Vasarely, who apply and explore certain physiological and perceptual effects of vision. I admire the ingenuity and enjoy the fun of these demonstrations, and I realise

23 Leonardo da Vinci, St. John the Baptist

24 Claude Lorrain, Landscape with Narcissus and Echo

25 Berghem, Mountainous Landscape

26 John Constable, Hampstead Heath with a Rainbow, *c.* 1828

27 W. Vandevelde, Gale at Sea

28 W. M. Turner, A Ship Aground

29 Backhuysen, Coast Scene

30 John Dillwyn Llewelyn, Photograph of a wave, taken at Brandy Cove, 24 May 1854

31 E. Manet,
The Melon

that the artists cannot be blamed if we find them somewhat marginal in importance. Claude Lorrain, to remind you of Constable's fine formulation, 'was legitimately connected with the chain of art', these painters cannot be, for there no longer is such a chain. It broke into disconnected links when the consensus broke down about the aims and functions of image making in our culture.

It is the lack of common purpose which also makes one hesitate to speak of experimentation in the sense which we observed in ancient Greece and again in the period reaching from the Renaissance to the late nineteenth century. For what I described as directed experimentation presupposes the existence of a problem for which good or better solutions can be offered.

There remains, then, the third possibility to which I briefly referred at the outset of this lecture, that rhythm of undirected trial and error that plays such a part both in organic and cultural evolution as natural selection leads to the elimination of misfits and the survival of the fittest.

Now this, if I understand it aright, is the tenor of the proposition advocated by the Bradford Trust. It puts 'forward for discussion the idea that an important criterion of fitness in cultures is in fact survival', and it proposes further that this survival may be legitimately compared to verification in science. 'Verification is here used in the sense of acceptance by a human audience though such acceptance may sometimes be delayed', a point to which I shall have to return.

I am not sure, as I have indicated before, that this interpretation fits art more than it fits other aspects of culture such as religious or social institutions; they must all be accepted in order to survive and what the historian tries to explain within the limits of his resources is precisely how and why a particular culture incorporated certain modes of life or styles of art. But despite this qualification I find this proposal particularly relevant to the situation of art today, and that, paradoxically, because if there is such a mechanism in art it has almost ceased to function in our time. We are familiar with such impediments to the 'survival of the fittest' in other fields than art. In politics the obsolete can cling to power by force. In economics the free play of the market can be distorted by protectionist tariffs or even by unscrupulous advertising. The forces which militate against the unfolding of art in modern society are more subtle and more insidious. They spring from a new kind of 'protectionism' unknown to previous ages. I refer to the belief in the verdict of the future which is also incorporated in the document before us. The passage I have in mind reads: 'The verification (of works of art) is of course not merely limited to our own view, or to that of contemporary opinion, but must be agreed or rejected by a large majority of those to whom it is submitted. Thus even if a work of art is 'ahead of its time' and generally rejected by contemporary opinion, future generations may recognise its truth and it may thus attain verification. On the other hand, it may simply come to be regarded as an irrelevant period piece'.

Now there is an interpretation of this statement which I fully endorse, as I have done earlier on when commenting on Cicero's anecdote about the poet who preferred the approval of Plato to that of a hundred thousand others. If you are sufficiently 'élitist' to assume that the percentage of discerning critics in any population is likely to be small, any good work has more chance of being recognised by such critics in the course of centuries than at any given moment. I suppose this is the process our document describes as 'averaging', but much also depends on accident, on who gets to know what work, just as the recognition of Mendel's experiment was long delayed because he published it in an out-of-the-way place.

But if I may remain with this illustration for a moment, we have also experienced in our time another and more sinister reason for the failure of Mendelianism. I refer to the situation under Stalin when it was unsafe to doubt the results of Lysenko's experiments purporting to disprove Mendel's laws. Whether culpably or not the experimenter was surrounded by a chorus of yes-men who impeded self-criticism and criticism.

In science such a stifling of the mechanism of negative feedback is still an exception. It is my contention that this is not so in art. For here the once innocuous belief in the verdict of the future has been turned into a weapon against any criticism, or 'fault finding'. Criticism is never pleasant and it is often uninformed. One can thus sympathise with the artist who stakes his faith in posterity, and that the more, the less he finds himself understood. One of Zola's heroes clings to this consolation as he defiantly confronts the hostile crowd at an exhibition with the battle cry: 'we have the verve and boldness, we are the future'. In this view rejection by the present is almost the guarantee of future fame, because, so the legend goes, all great artists, indeed all true geniuses were derided by their contemporaries. That this is bad history goes almost without saying. What matters more is that it is also bad philosophy. For the belief that good art not only may, but must be 'ahead of its time' is not rational as is the faith in 'averaging'. It is rather part and parcel of that philosophy which Popper has criticised as 'historicism', the belief that there is a law of progress in history which it is not only futile but actually wicked to resist.[19] It is wicked, because whatever sufferings may be caused by revolutions, wars and massacres they are merely the inevitable accompaniments, the birthpangs of a better and brighter age. It is a philosophy which would hardly have been accepted by so many if it had not carried the consolations of religion into the arena of politics, promising victory to its adherents and damnation to its opponents. Transferred to the sideshow of art this ideology of inevitable progress is of course responsible for the notion of the avant garde, those advance parties of art which will plant the banner of the new age on the territory which will be settled by the next generation of artists. I find it noxious, because it really abolishes the very belief in values which the other interpretation upholds. There is no bad art and good art, only antiquated and advanced art. One day the scales will drop from the eyes of the purblind majority and they will accept what they now deride, but when this happens it is time for the avant garde to get busy and court martyrdom for the sake of the next age.

Now it stands to reason that this version of 'futurism' can lead to a crippling elimination of all negative feedback, and with its disappearance the notion of an artistic experiment also loses its meaning. However the artist chooses to paint his bunch of carrots none of his contemporaries can judge it. What is worse, it becomes doubtful that he can judge it himself. Our standards, our conscience, moral or artistic, are derived from our environment. We are free to criticise and modify them but without criteria of what is good and what is better we cannot submit our ideas to the judgement of our 'corporeal eyes'. It is this unintended breakdown of standards that has made it so hazardous to compare the life of art today with the life of science.

For contrary to the rigorous standards by which any scientific paper is judged when it is submitted for publication the lack of criteria has led to a loss of nerve among critics and, what is more serious, it has also intimidated the public. Intimidated precisely because their self-respect is threatened. If they fail to acknowledge the art of the future they show themselves to be mentally backward. I am not the first to recall in this situation Andersen's famous parable of the Emperor's new clothes which are sold to His Majesty by cunning merchants, who claim that the new material has the added advantage of only being visible to those who are fit for their office. In the end it takes the innocent child to call out 'but he has nothing on'. We cannot be helped by such innocence, for how could a mere child discern the art of the future?

The story is germane to my subject because it is of course the story of a psychological experiment. All hoaxes can be so interpreted, and what hoaxes in the field of art tend to show is the degree to which acceptance or rejection is influenced by our suggestibility. Vasari tells us of such an experiment which Michelangelo made while at work on his famous David. When the communal leader of Florence, the gonfaloniere Piero Soderini came to inspect it, he praised the statue but thought that the nose was too big. Whereupon Michelangelo climbed up on his scaffolding, took a few grains of marble lying around and pretended to chisel off a piece of the nose, letting the dust fall on Soderini, who promptly said 'now I like it better, you have given it life'. Can anyone be sure he would not have fallen for the trick? And can we be surprised that Apelles preferred the incorruptible horse to the corruptible judgements of human beings? The trouble is only that what is corruption to some, is conversion to others.

When George Braque, in 1906 or 1907, first saw the revolutionary composition by his friend Picasso which is called rather squeamishly *Les Demoiselles d'Avignon* – for it represents a group of prostitutes in a brothel – he is reported to have said, 'it is as if you asked us to drink petrol'. But art is an acquired taste and I confess that actually I have come to like drinking this petrol – the so-called experiment of cubism, which started with this picture, had led to the discovery of interesting and amusing visual puns and ambiguities which I enjoy as long as I am not told that they correspond to Einstein.

Art is not science and the art exhibition differs in more than one respect from the laboratory where the results of experiments are soberly assessed and repeated by the investigator's peers who will find the faults, eliminate error and work in the direction of progress. I do not want to be misunderstood, I do not want to suggest that in contrast to science the world of art today is entirely governed by brainwashing and bandwaggons, but I do think that our art is not safe from corruption from these forces and that we must be aware of these distorting influences if we are to take the comparison of works of art with scientific experiments further. What we call art, as I have stressed, has served a variety of functions in a variety of cultures. If today we see artists trying out and trying on an unprecedented variety of modes, media and effects we should enjoy the opportunity of accepting or rejecting these mutations without a feeling of guilt or pride. I believe that the young have largely come round to this point of view. Indeed this reaction against portentousness is the nugget of value I discern in my more optimistic moments in the noisy carnival of anti-art; if only portentousness did not so often catch up with it, as it has caught up with 'Dada'! Not that I would ever wish to discourage solemn emotions in front of transcending masterpieces. I fully agree with the Bradford document that there exist such works which, as the saying is, have 'stood the test of time'[20] and I share the feelings of gratitude and admiration for these supreme achievements no less than for the intellectual conquests of science. But while science is a coherent body to which every practitioner can make a major or minor contribution, art is not. It has its valleys and plains as well as its soaring peaks. If we equate the artist with the scientist by regarding both as equal servants of human progress we may tempt our art students to see their role as that of prophets and oracles with disastrous results for their mental equilibrium and their capacity for self-criticism.

It was not that which Constable meant when he addressed an audience which included Faraday. If you listen sharply you can perhaps still hear his final plea for humility. 'It appears to me that pictures have been overvalued, held up by a blind admiration as ideal things . . . and this false estimate has been sanctioned by extravagant epithets that have been applied to painters as "the divine", "the inspired" and so forth . . . yet the most sublime productions of the pencil are . . . the result, not of inspiration, but of long and patient study, under the direction of much good sense'.

24th February, 1978

References

1. Constable's notes for his lectures were published in C.R. Leslie, *Memoirs of the Life of John Constable*, London, (1843) I quote from the edition by Jonathan Mane, p. 323, London. (1951).
2. H.A. Groenewegen-Frankfort, *Arrest and Movement*, London, (1951) offers a very negative assessment of the Amarna style.
3. Brian Vickers, *Towards Greek Tragedy*, Ch. 10, Four Electra plays, London, (1973).
4. *cf.* my *Art and Illusion*, Ch. 4. New York and London. (1960).
5. For this and the following see also my chapters on the Renaissance Conception of Artistic Progress in *Norm and Form*, London (1966) and From the Revival of Letters to the Reform of the Arts and The Leaven of Criticism in Renaissance Art in *The Heritage of Apelles*, Oxford, (1976).
6. Leonardo da Vinci, *Treatise on Painting*, ed. A.P. McMahon, vol. I, p. 161, (amended). Princeton, (1956).
7. *The Heritage of Apelles*, pp. 38–75 as quoted above.
8. *ed. cit.* p. 305.
9. Karl Popper, On the Sources of Knowledge and of Ignorance, *Conjectures and Refutations*, London. (1963).
10. *The Works of John Ruskin*, ed. E.T. Cook and A. Wederburn, XXII, p. 209. London (1903–1912).
11. W. Holman Hunt, *Pre-Raphaelitism and the Pre-Raphaelite Brotherhood*, Ch. VI. London. (1905).
12. Anita Brookner, *The Genius of the Future*, London. (1971)
13. My quotations are taken from section II of the novel.
14. Sven Lövgren, *The Genesis of Modernism*, pp. 71 ff. Stockholm. (1959).
15. R.A. Weale, The Tragedy of Pointillism, *Palette*, pp. 17–23. (Sandoz Ltd.) Basle. (1972).
16. Quoted after H.R. Rookmaaker, *Synthesist Art Theories*, p. 1. Amsterdam. (1959).
17. Renato Poggioli, *Teoria dell'Arte d'Avanguardia*, Ch. VII. Bologna (1962).
18. Sixten Ringbom, *The Sounding Cosmos*, p. 33 ff. Abo, (1970) traces the history of this notion.
19. K.R. Popper, *The Poverty of Historicism*, London, (1957), see also my lecture, Hegel und die Kunstgeschichte, *Neue Rundschau*, 88/2, (Spring 1977) English translation in *Architectural Design* No. 51/6/7, pp. 3–9. (1981).
20. *cf.* my Romanes Lecture, (1974), *Art History and the Social Sciences*, in *Ideals and Idols*, Oxford. (1979).

APPENDIX

A Symposium on Scientific Method and the Arts

Following the Richard Bradford lectures, Lord Briggs chaired a Symposium on Scientific Method and the Arts at the Ciba Conference Centre in London on 24–26 May 1978. Invitations were sent to all the former lecturers and others who had shown interest in the Trust. Some were not able to come and the final membership consisted of three former lecturers, Professor Kitto, Professor Wickham and Jacquetta Hawkes, together with Professor John Hale (Department of Italian Studies University Collge, London, and Chairman Board of Trustees National Gallery), Kenneth Lindsay (formerly Minister of State for Education and a Founder Member of PEP), Dr Hans Lissmann (Zoological Laboratory, Cambridge), Sir Ralph Verney (Chairman British Association Working Party on Science and the Quality of Life), Mrs Kitto (representing the study of music) with the Trustees of the Richard Bradford Trust: Professor G.W.A. Dick, Dr A.M. Stewart-Wallace, Dr R.B. McConnell, Dr R.H. McConnell.

Lord Briggs opened the meeting by explaining that the purpose was first to discuss the aims of the Richard Bradford Trust, which are very directly associated with an exploration of the relationships between the methods of scientific investigation and artistic creation and an understanding of the map of learning; secondly to give some help to the Trustees, as to how their aims could be further pursued.

Each member had been asked to prepare a short talk on his or her particular subject, abstracts of which were circulated before the meeting; these are presented in this Appendix.

Valuable discussions followed all the talks and many excellent ideas were put forward. It is regrettable that limitations of space forbid paraphrasing them all, but a verbatim report was made and multigraphed which is available on request from the Richard Bradford Trust, Streatwick, Streat, Hassocks, Sussex, BN6 8RT, to persons who would be especially interested.

We were especially grateful to Lord Briggs, who wound up our final session of discussions with much excellent advice as to the further implementation of the Trust, and to Mr Kenneth Lindsay for his description of the foundation of PEP, which could well serve as a model for further development of the Trust.

Abstracts of Talks

1. R.B. McConnell: *The Aims of the Richard Bradford Trust*

In my Introduction (see pp. 1–25 above) I have described how many years ago I developed certain ideas based on my scientific experience and general reading, which I hoped to work out after my retirement, but realisation of my ignorance and inadequacy showed me the necessity of consulting other sources of learning and experience and this eventually led to this meeting.

It is a fundamental assumption that all creatures have urgent needs such as food and warmth for the preservation of life and survival of their kind, and reassurance when darkness seems to fall. Mankind only is driven by spiritual necessities to seek meaning in life, largely through what we call 'religion', although this word has of course had many meanings through prehistory to the present day. In religion a belief in some master idea or mode of life is generally required, but as generations pass a tendency grows to seek a basis for such belief.

Western society is basically a Greco-Roman-Judaic culture, and even if we do not belong to a formal religion we obey laws which are based on an ethic close to that of the New Testament. However, when we start asking *why* we should behave in this way the standard answer has been: 'because God tells us to'. If therefore, we start to wonder who God is and what the evidence is for His existence, we are likely to wonder also if our ethical values are soundly based, and if there is any means of testing them. In my Introduction I have suggested that our civilisation is now in this position, our religions thus no longer command absolute obedience and our whole ethical stance is therefore in question.

Speaking generally, mankind's needs may be said to fall into two baskets. One for spiritual and ethical values and the other for material needs. The first has long called forth religion, philosophy and art; the second, technology. The latter was for many millennia satisfied by observations such as that certain foods nourish while others poison, fire warms and so forth: but in the course of time minds began to ask the question 'Why'.

In prehistory and antiquity the puzzles of nature were answered by simple statements of belief, occasionally backed by experiments such as Archimedes overflowing his bath. But eventually, as the definition of factual evidence became more refined, experiment became more controlled and science as we know it was born. This led directly to the verification or falsification of ideas and to the establishment of relative truth within a defined frame of reference. When dealing with simple matters such as the expansion of a metal bar with heat, the results can be quantitative and strictly repeatable. In a science such as geology, on the other hand, a large number of variables are frequently involved and the results are only accepted as relatively correct within varying degrees of accuracy.

When we turn to the first basket, however, is there any method of investigation which can give us results of acceptable value? Formal religions and philosophies have all aimed at 'truth', but how many of their arguments and final statements in the past have not been queried and often superseded? And how many of ours will not go the same way? Are there any criteria which could prove or disprove such statements even within very wide limits? I have in my Introduction put forward for discussion a possible answer to this question.

Statements referring to subjects in basket one are made by artists in music, poetry, drama, the visual and other arts, which I suggest *can* be tested and either verified or falsified in the medium to which they refer. Art is a product of the human mind and hence its products must be tested in this frame of reference. As we have seen that the scientific method applied to questions in basket two can deal even with large numbers of variables by averaging them out, so to investigate subjects in basket one we have to find a method which can average out a truly vast number of variables. I suggest that art is such a method.

As the scientist develops a hypothesis and then tests it within its frame of reference, cannot a work of art also be tested within its *appropriate* frame of reference, namely by exposure to a human audience, not considering the opinion of one person only, no matter how eminent, but taking a wide average not only of people but of generations? As extreme examples, the visions of Homer, Michelangelo and Beethoven have been tested over the generations and still ring true today.

If indeed there is anything in what I am saying, then, as the scientific method deals with relative verification and falsification of material moment and hence improves the material welfare of society, so art can lead to verification or falsification of ethical concepts and so discover directions in which our evolving culture should try to move.

2. H.W. Lissmann: *Imprinting*

For purposes of analysis Dr McConnell has suggested that the material and spiritual needs of humanity should be put into two baskets. A third basic assumption is that man has evolved from other creatures, and perhaps one should ask the question why man alone is a 'two-basket' creature. Why should he occupy such a unique position, and how did he get saddled with his philosophical intellect, this urge to search for a meaning in life? If you want to explore and influence human progress it seems important to examine first the two foundations upon which our reactions to the outside world rest, i.e. evolutionary history and individual experience.

Perhaps I might add one small point about observation from the moment a baby's eyes open and storage begins of information in the memory, on which both artists and scientists alike draw with advance of years. Recent work has shown how very important experiences during an early critical period can be to future outlook and behaviour. A very rapid learning process takes place in young creatures, called imprinting. For visual stimuli, for instance, it has been shown through radioactive tracer methods that important biochemical changes take place in the visual brain centres during imprinting. It has also been shown by Colin Blakemore, the recent Reith Lecturer, that when kittens are brought up in large dustbins with vertical black and white stripes, the visual fields in the visual cortex of their brain are all orientated in one and the same direction, and in later life such cats manoeuvre quite happily between a forest of vertical pallisades, but not between horizontal bars. The reverse happens when kittens are confined during the critical imprinting period in dustbins with horizontal stripes. Similar events take place in the brains of chicks when they hatch and quickly learn to follow their mother. Emerging from an incubator, they can be imprinted onto almost any object, a flashing light, a block of wood, or onto man himself. So we are beginning to learn what histological changes are taking place in the brain during this process. This makes one wonder what may be the critical period for exposure to a Greek tragedy, or Western civilisation, to produce lasting effects and a lasting preference.

3. H.D.F. Kitto: *Greek Tragic Drama*

In Athens, during the 5th and 4th Centuries B.C., tragic drama was an immensely popular art, composed not for an *élite* but for the multitude, though by the 4th Century it had turned into something much less impressive. Why did that happen? In our own times, say the last hundred and fifty years, interpretation of the thirty 5th Century plays which have survived, or of most of them, has been so diverse, even contradictory, as to give the impression that modern scholars are pretty much at sea with them. What is the reason for that? The chief reason lurks in the phrase 'religious drama', which is freely used in describing it. And why not? The plays were written, exclusively, for a religious occasion, the annual Festival of Dionysus and are themselves full of references to a god or gods, who sometimes appear, as actors, on the stage – though not very often Dionysus himself.

But it is an interesting fact that the classical Greek language did not possess a word equivalent to 'religion'. Instead, it made do, quite successfully, with a phrase like 'what pertains to the *theoi*', and the word *theos* we translate as 'god', since our language gives us no alternative. Very often the word 'god' serves well enough, but just as often it is misleading – like related words such as 'divine', 'religion'. For this reason: that although the word *theos* was freely used in the 'religious', transcendental sense with which we are familiar, it was also used of any power or principle which demonstrably, and infallibly, operates in the universe, physical or moral – of any fundamental fact of life. Thus, facts like Love, Hatred, Wrath, could be and often were, called *theoi*. That explains why the Greeks perceived so many of them.

This explains also why the action of a play is so often presented, concurrently, on two levels at once, the human and the 'divine' – in a kind of double perspective. Thus, in the Oresteia, Apollo commands Orestes to avenge his father and Orestes gives several 'commanding' reasons why he would have to do it in any case, even if Apollo had not spoken. Such duplication is foreign to our own religious tradition. We are apt to think that the omnipotent god *makes* the man do it; but no: regularly the man is acting quite independently. The fact that a *theos* is (as it were) acting alongside of him is the imaginative Greek way of suggesting that in the spontaneous human action some fundamental fact of life is operating. To adapt a phrase of Aristotle's, the action is both what *did* happen and what *would* happen.

It seems to have been instinctive in the Greeks to seek the general in, or behind, the particular. Is that the reason why they contributed more than any other ancient peoples to the development of modern science?

4. Jacquetta Hawkes: *The Archaeological Approach*

Perhaps the greatest value of the archaeological approach to our problems is that it reveals artistic creation in the days of its intellectual innocence: pure imaginative art 20,000 years before the Ionian dawn of scientific thought. Palaeolithic art was deeply involved with rites and religious feelings and like them was spontaneous, intuitive, rooted in the unconscious.

As the earliest known art it quite simply *is* part of human progress: it can be said to be 'verified' by the approval of its makers and ourselves. Yet its quality had nothing to do with social or moral codes. Art preceded such codes. Plato's view was that poetry was the natural food of the very young.

Archaeology has also enabled us to observe the rise and disappearance of whole civilisations and to appreciate the unique 'form' of each. Past civilisations remind us of the historically normal importance of the imaginative arts, their involvement with almost all aspects of life. Their nature also appears to reveal something of the essential being of the creators: for example when contrasting Minoan and Aztec art forms.

These facts are sympathetic to Bradfordian thought. On the other hand the fact that nothing seems able to avert the decadence and death of cultures and that it is technical advances that survive must be significant. In this sense art is short and technology long.

As 'verifiers' what is our verdict on good works of art that are decadent, or that glorify war and cruelty?

From the time of 'the Greek miracle' with its bequest of philosophical and ethical ideas, we have artists of genius inheriting, with the individualism of Western civilisation, a great intellectual and moral tradition. We get a supreme combination in the masterworks often named by Dr McConnell. Surely the two are distinct, and when it is said 'art can lead to verification or falsification of ethical concepts' such judgment is equally distinct and comes from moral philosophy and history, not from aesthetics. The contribution of the artist is in making the concepts palatable, moving, enduring.

5. Glynne Wickham: *A Sense of Occasion*

'A sense of occasion' is something that may be said to be universally experienced by human beings regardless of race, colour, or phase of social development. Yet it is scarcely subject to scientific measurement by altimeters, clapometers or any other such instrument that is mathematically precise in its verdict.

This paradox arises neverthless from an innate desire and determination on the part of all primitive societies (which is also sustained by more sophisticated societies) if not to measure, then at least to distinguish, moments or events that they believe to be important in life from those that can be dismissed as trivial; things they regard as significant from those that are not; the extraordinary from the ordinary; the supernatural from the natural. This phenomenon is recognisable wherever and whenever a community elects to dramatise an occasion or event; and by that I mean to call a halt to time in order both to dress the event in festive gear and call attention to it, and to comment on it. Celebration and commemoration thus combine within ritual to create a sense of occasion.

The Romans, in the wake of the Greeks, coined the word *feriae* to describe these occasions and the word *ludi* to describe the festivities attached to them; the latter they subdivided into two categories, athletic and mimetic; and in time their mimetic games came to embrace *joci*, or word games. These distinctions were never complete since the presentation of the *ludi circensis* was endowed with a high degree of theatricality, and the *ludi scenici* were frequently structured on combat, whether physical or verbal.

Christian Europe inherited this pattern of marking 'occasion' with public holidays and of dramatising the events celebrated. In doing so it preserved a remarkable proximity in dates between its own 'holy days' and those of the Roman, Celtic, Teutonic and other earlier cultures which Christianity superseded. Ample evidence of a strictly factual kind survives to prove that the Roman Catholic Church elected, in its own drama between the 10th Century and the Protestant Reformation, not only to preserve the 'holy days' that it found *in situ*, but to endow them with a specifically Christian significance. Again, however, the purpose of the resulting drama, at first in Latin and subsequently in vernacular languages, remained to commemorate, celebrate, and explain. Its function therefore was akin to measurement in that it sought first to separate the 'Red Letter' days of the Calender from the prevailing black ones – the wheat from the chaff, as it were – and then to particularise, in didactic fashion, the relationship of each day selected for capitalisation in red to the Church's own dogmas. It would thus be meaningless, self-defeating, to present the *Visitatio Sepulchri* particular to Easter Sunday on Christmas Day, or the *Visitatio Praesipe* on Easter Day, or indeed to try to merge the two.

What the historian (or the social anthropologist) is here confronted with is society's way of both measuring and responding to its own emotional needs. On both counts the art of drama thus offers an intuitive pragmatic resolution of the paradox described in the first paragraph of this Abstract rather than a conceptualist one. When a society forgets this and seeks instead to make its drama fulfil strictly rationalist or hedonistic ends, the drama will itself collapse by failing to find an audience, as our own is in acute danger of doing today.

6. John Hale: *Art and the Human Condition*

For art to play a role in 'tending to increase the stature of humanity in the world', the audience that can derive ready and thoughtful delight from it must be greatly extended.

Great art (for it is the response to *that* that must be fostered) represents the farthest expansion of faculties latent in everyone: making things with the hand; seeing – (the eye, also, is genetically an instrument for survival); arranging shapes and colours into patterns – the 'hand', 'eye' and 'symmetry' that created Blake's Tiger. In many schools these are already catered for, but in the limited cause of 'self-expression' – limited because only a very few can express themselves fully by visual means and because self-expression invites the eye to turn inward rather than to man and nature and the art of the past – and then only for a few years before art is submerged in exams. Art education for the majority stops before the necessary next stage is reached: appreciation of how paintings can transmit a marvellously rich package of responses to the fact of being alive, responses that should be as unresentedly exhilarating as is the ballet dancer's grace compared with our own clumsy leap over a ditch. And it stops before solving its major problem: that paintings are *still* while practically everything else we do involves movement: reading, we turn the pages; music flows past as do the words on the radio or the images in a film or on television. It has become positively unnatural to look attentively for some time at an unmoving object except for lovers, birdwatchers and astronomers. It is something that has to be taught.

How can schools, the media, as well as museums and galleries, further encourage a sense of the naturalness of turning to great works of art as a fusion and enrichment of organic impulses all are potentially equipped to share?

It may be that there are links to be found between the methods and manners of arts and science. Certainly, it is interesting to speculate about them. But the debate will be merely esoteric unless it reaches larger numbers than exist at present of those who are aware of the significance to humanity of masters who are the extension points of our visual sensibility, unless looking comes before linkage, what is specific to art before what is generic to society. Perhaps this is a theme the Trust might investigate on a future occasion?

7. Arthur Stewart-Wallace: *Some Observations on the Anatomy of the Human Intellect*

The division of man's needs into two categories, the spiritual and ethical on the one hand and the material on the other, seems to me as a neurologist to have some basis in the way the human brain is structured and functions.

While our emotional mental processes are localised to a large extent in the older more primitive limbic system which we share with our mammalian forebears, the most characteristically human activities, which include mathematics, science, technology, music and the arts, are seated in the more recent neo-cortex which has developed to such an enormous and unique extent in man and makes up the two cerebral hemispheres.

It is widely known that language in the majority of people is a function of the left hemisphere and a stroke affecting the left hemisphere may cause loss of the ability to speak, read and calculate. It is not so well known that a right hemisphere stroke may produce impairment of pattern recognition, three-dimensional vision, musical ability and thinking in a holistic way. This and other evidence show that there is a highly sophisticated ability to perceive and think which bypasses verbal and analytical processes. We term this *intuition* and it is involved when, for example, we recognise at a glance a face or recall a piece of music. Intuitive thought is mainly a function of the right hemisphere, our awareness of which the brilliant and dramatic verbal abilities of the predominant left hemisphere tend to obscure.

Effective understanding requires both the intuitive search for patterns in nature of the right hemisphere and the critical analysis of the left. The result on behaviour of these two different cognitive functions is exemplified by the human response to the sight of blood. There is the practical scientific reaction of the left hemisphere leading to effective treatment and the intuitive reaction of the right hemisphere causing a feeling of distaste and a vicarious sympathy which motivates us to take steps to avoid a repetition. Both reactions have a survival value for the species.

When D.H. Lawrence said 'It is no use telling me that the moon is a dead rock in the sky', he was expressing his awareness that if we leave out our intuitive appreciation of its beauty and romantic associations we miss half its meaning.

Critical thinking without creative or intuitive insight is sterile. It is necessary that the nature and direction of our rational and analytical endeavours are influenced by their ultimate human implications as revealed by intuitive thinking. Even when scientific insights are associated with right hemisphere intuitions they can be adequately proved or disproved by left hemisphere analytical processes. In creative art, however, which has major right hemisphere intuitive components, verbal arguments on its validity are invariably inconclusive.

The Bradfordian proposal to use as a test of the verification or falsification of artistic subjects the average of the opinions of generations of human beings must involve a large measure of right hemisphere intuition as well as left hemisphere analysis. Such a method of assessment must therefore necessitate that collaborative action of both hemispheres essential for our fullest understanding, and is made the more acceptable by what we know of the structure and functions of the human brain.

At this meeting when McConnell says about appreciation of art 'it rings a bell', Jacquetta Hawkes speaks of spontaneous intuitive features in Palaeolothic art, Professor Kitto of the intuitive feelings for the general in the Greeks, John Hale of how paintings can transmit a marvellously rich package of responses to the fact of being alive, Kenneth Lindsay of knowing quality when we meet it, they are all implicitly indicating that intuitive rather than rational thinking predominates in creative art.

In these necessarily brief and over-simplified observations, I have not elaborated on the less conscious but important emotional feelings for art, nor on altruistic or religious behaviour, the beginnings of which we perhaps see in mammals and birds in their devotion to the care of their young and the attachment of domestic animals to man. These are largely associated with the limbic system of the brain. Nor have I dealt with the even older hindbrain, the most developed part of the brain in reptiles and insects. The hindbrain is probably the main seat of some of our own hierarchical and ritualistic behaviour, which is such a striking feature of many forms of animal life.

[One very apposite contribution to the general discussion was that made by Professor Glynne Wickham when he commented that Shakespeare had actually written a play –

Love's Labour's Lost – which illustrated Dr Stewart-Wallace's paper, showing how the intuitive personalities, Costard and Jaquenetta, won out over the King of Navarre and the other serious-minded characters.]

8. Kenneth Lindsay: *Freedom, Quality and Fraternity*

My interests, as a political scientist, as an amateur sociologist and one involved in education, appear to have little place in the Bradford syndrome. But my concern for freedom, quality and the application of the scientific method to social problems is central. Perhaps freedom needs re-defining every generation; I prefer that of Graham Wallas – 'the opportunity for continuous initiative'. This definition would be woefully incomplete as it stands, because the initiative might be murder or anything else. Quality is another matter, although unmeasurable; we know it when we meet it, whether in a play by Euripides or Robert Bolt or whether in political diagnosis.

In 1931, a critical year in history, I was engaged in a joint venture to discuss some first principles for social action, resulting in the founding of PEP. The name PEP, stood for Politics, Economics and Planning. We were very careful to have much more interest in freedom than planning, but we did want a framework. PEP was created by events. In 1931 there were about three million unemployed and there had been a bankruptcy in political thinking. We wanted to change things. We were not sure about the Good Life, but we did believe in the accumulation of facts. I cannot tell you how careful the accumulation of facts was and how imaginative as well as 'assiduous'. Oliver Roskill and Max Nicholson and others saw to that. There was research for two years before any publication. We were worried about the critical state of the country, as I think a lot of people are today.

We gathered together a number of specialists, but specialists of a particular order, who were capable of seeing their specialism in a wider context, if not *sub specie aeternitatis*.

Fraternity was there; PEP became a living chain of men and women with a will and a purpose. It is not necessarily relevant that we anticipated several institutional reforms; they may be ephemeral. It is possibly relevant that, following in the footsteps of Haldane, Graham Wallas and Seebohm Rowntree, we conceived an economy which would be a mixture of free, voluntary activity and social goals. These were to be basic principles in every sphere of action – political, economic, educational.

The Bradford initiative has coincided with measured reconsideration of our re-written Constitution and of the British philosophy of administration. We seem to have lost both direction and purpose, resulting in less freedom and lower quality. I feel sure that the application of scientific method can help to recover our wayward sense of direction, but something more basic is needed to restore lost values.

It is one thing to direct the scientific method to new problems (Karl Popper's image of the searchlight scanning a night sky for planes is apposite); the whole scientific enterprise can within limits be expanded, but not the scientific method itself. Perhaps the non-scientific method applied to man and the world can give us meaning, purpose and a vision in which everything coheres. There can be no science of the Good Life; this is the area for philosophers, mystics and prophets.

9. Ralph Verney: *The British Association's Working Party on the Quality of Life*

The British Association, among its many other activities, tries from time to time to explore a number of topics where science and technology affect society for good or ill and to bring the results of these investigations to the attention of scientists and the general public. Such topics have included the social consequences of new developments in genetic engineering, science and ethics, and more recently the impact of science and technology on the quality of life in the industry-based society of Great Britain. We have been working on this difficult subject for over a year; we hope to publish our conclusions in the form of a book.

It is very difficult to define what we mean by the good life: a monk, a pop musician and a housewife would give different answers and there is likely to be a basic cleavage between those who are concerned with the cradle to the grave and those whose horizon is baptism to eternity. Industrialists and conservationists, conservatives and socialists, bosses and workers seem to hold opposing views – to identify the good life with a more efficient, a cleaner or a collectivist economy. But the ultimate goal is the same, to make ordinary people better off. The disagreements are largely about the means rather than the ends. Definitions of liberty and equality vary widely, but as Halsey said in his Reith Lectures, the catalyst is fraternity and fraternity tends to be overlaid in our Society by politics, the sociology by economics, by the god of growth. So we are coming to the conclusion that one of the most important elements in the quality of life is 'esteem' both of your peers and of yourself. We have identified the aspects which we wish to explore in depth, because it is in these spheres that the impact of science and technology is most immediate and we have discussed these fields with a number of experts. The subjects we cover are Health, Home and the Family, the Quality of working life, Leisure, the Physical Environment (including housing and transport) and Communication. We regard these as the spheres in which science and technology makes its greatest impact on the quality of life in our urban-based society. There are conflicting values in all of them and in all art has a place, however precarious: but that is a subject we have omitted from our terms of reference, though it will be implicit in our conclusions, because creativity is an obvious and essential ingredient of the good life.

We are quite clear that the quality of life and the standard of living are not the same thing, that science must submit to ethical disciplines and technology to ecological constraints, that conservation is as important as growth to the good life and that job satisfaction is not simply a matter of more money.

Index

Index